Mastering Logarithms

Aim

This book is designed to help secondary school students (Grades 10-12) systematically master the laws and principles of logarithms. With more challenging concepts introduced progressively each year, it serves as a long-term resource that students can revisit as they advance in their mathematical studies. Perfect for IGCSE, IB, SAT, ACT, and international mathematics programs.

Content

This book covers the syllabus content on logarithms required at all stages and levels of secondary school. The initial sections introduce elementary rules, which develop in complexity with the introduction of additional laws and strategies. The extension section includes challenging questions which develop problem-solving strategies, extend capable students and prepare them for senior school exams and mathematics competitions. In the final stage of the workbook, students will master the strategies required to perform graphical transformations of logarithmic functions, dilations, translations and reflection. Although focussed on logarithms, the book also requires a basic understanding of algebraic techniques. It is suitable for all secondary students and caters to a wide range of abilities. Depending on their ability, students can progress systematically year by year thereby strengthening their mathematical understanding of this topic.

Why is this book different?

- Worksheets are carefully graded in difficulty, allowing students to progress at their own pace.
- Concepts and strategies are introduced in small, manageable steps, reinforced with worked examples.
- This book encourages independent learning while systematically building mathematical understanding.
- Challenging competition-level questions provide enrichment opportunities for advanced students.

CATERING FOR ADVANCED STUDENTS

The final sections of the book, including review and extension exercises, provide challenging problems for advanced learners and senior students. These problems require deeper mathematical insight, logical reasoning, and problem solving finesse. Fully worked or guided solutions are provided to support independent learning.

WHY TEACHERS AND STUDENTS NEED THIS BOOK?

This book is an invaluable classroom resource, offering structured self-guided exercises that allow students to progress at their own pace while giving teachers more time to provide targeted support. It supplements standard textbooks and helps students master logarithms efficiently, reinforcing their foundation in mathematical skills.

This book is perfect for homeschooling families, self-study, tutoring, or classroom use worldwide

ABOUT THE SERIES

This workbook is part of a new series co-authored by Anne Joshua and Heidi Postle and marks a new collaboration between Heidi and Anne, combining their expertise to create a structured, progressive approach to mastering indices (exponents), surds (radicals) and logarithms. Anne Joshua is the author of the successful Mastering HSC Mathematics, Enrichment Maths for Secondary School Students, and the Enrich-e-matics series (Books 1–6 for primary students).

Each book in this series focuses on a specific mathematical topic, providing carefully graded exercises and comprehensive study material that supports independent learning and exam preparation. By systematically building on key concepts, this series ensures students develop a deep and lasting understanding of mathematics.

- **Mastering Indices**
- **Mastering Surds**
- **Mastering Logarithms**

GUIDED/WORKED SOLUTIONS

All questions have answers, and the more challenging and review questions include guided solutions to support independent learning and deepen understanding.

Grade Level Placement

This workbook is designed for secondary students across multiple international curricula:

- **IGCSE:** Extended Mathematics and Additional Mathematics (ages 14–16)
- **IB MYP:** Years 4–5 in Extended Mathematics (ages 14–16)
- **IB DP:** Standard Level and Higher Level (ages 16–18)
- **US Common Core:** Algebra 2 (typically Grade 10–11, ages 15–17)
- **Australian Curriculum:** Years 10–12

The progressive structure allows students to work at their own pace, making it suitable for various year levels within these ranges.

Copyright

Reproduction and communication for educational purposes

National Library of Australia

ISBN 978-1-7640499-2-4
PO BOX 6203, Malabar,NSW 2036 AUSTRALIA
masteringmathsseries@gmail.com
www.masteringmathsseries.com.au

First Edition, published 2026 by Anne Joshua and Heidi Postle

ACKNOWLEDGEMENTS

From Anne:
My husband Doug for his love, unconditional support and encouragement. My wonderful family for their love and the happiness they bring to me.

From Heidi:
Thank you Oshi, for your unwavering support, patience, and love. To my children, thank you for your help and suggestions. And to my mother, thank you for keeping me company through so many late nights of editing.

— *Anne Joshua and Heidi Postle*

Discover More Titles by Anne Joshua

- Mastering HSC Mathematics Year 11 Advanced
- Mastering HSC Mathematics Year 12 Advanced
- Mastering HSC Mathematics Year 11 Extension 1
- Mastering HSC Mathematics Year 12 Extension 1

These above books are a rich, comprehensive resource for students and teachers (Jonathan Lee Co-author)

- Enrichment Maths for Secondary School Students – Developed to challenge and extend capable learners
- Enrich-e-matics Series (Books 1–6) – Fun, challenging problems for able students for Years 1–6
- Enrich-e-matics Teacher's Book

These resources have helped thousands of students to strengthen their skills and confidence in mathematics and provided an excellent resource for motivated and able students

Logarithms

Introduction

The logarithm of a number is the power (or index) to which the base must be raised to give that number.

Definition: If $\log_a y = x$, we say that the logarithm of y to the base a is x since $a^x = y$.

Note:

1. The base is written as a subscript to the word 'log' as $\log_a$
2. $a > 0$. i.e. the base is always positive, $a \neq 1$
3. $\log_a x$ is only defined for $x > 0$

Example 1

$\log_{10} 1000 = 3$ since $10^3 = 1000$

Example 2

$\log_2 8 = 3$ since $2^3 = 8$

Exercise: Write equivalent index statements to justify the following log statements

1.	$\log_{10} 100 = 2$	2.	$\log_{10} 10 = 1$
3.	$\log_4 16 = 2$	4.	$\log_2 32 = 5$
5.	$\log_2 16 = 4$	6.	$\log_5 1 = 0$
7.	$\log_{12} 1 = 0$	8.	$\log_2 x = y$
9.	$\log_a y = x$	10.	$\log_y x = a$

Rewrite each of these expressions in logarithmic form.

Example 3

power

base ↘ ↓ ↙ number

$$4^2 = 16$$

So, $\log_4 16 = 2$

Example 4

$$3^{-1} = \frac{1}{3}$$

So, $\log_3 \frac{1}{3} = -1$

Exercise: Write equivalent log statements to the following

1.	$2^4 = 16$	2.	$3^2 = 9$
3.	$5^0 = 1$	4.	$4^3 = 64$
5.	$10^3 = 1000$	6.	$10^1 = 10$
7.	$7^1 = 7$	8.	$7^2 = 49$
9.	$7^{-1} = \frac{1}{7}$	10.	$1 = 9^0$
11.	$\sqrt{5} = 5^{\frac{1}{2}}$	12.	$\frac{1}{9} = 3^{-2}$
13.	$\frac{1}{10} = 10^{-1}$	14.	$5^{-2} = \frac{1}{25}$

Sometimes you have to evaluate an expression.

Example 5

Evaluate: $\log_2 16$

Solutions:

$$\begin{aligned} \textit{Let} \quad \log_2 16 &= x \\ 2^x &= 16 \\ 2^x &= 2^4 \\ x &= 4 \end{aligned}$$

So, $\log_2 16 = 4$

Exercise: Evaluate the following

1.	$\log_2 4$	2.	$\log_2 8$
3.	$\log_2 32$	4.	$\log_3 27$
5.	$\log_3 81$	6.	$\log_4 64$
7.	$\log_6 216$	8.	$\log_2 128$
9.	$\log_{2\sqrt{2}} 4$	10.	$\log_{\sqrt{10}} 100$
11.	$\log_{3\sqrt{3}} 9$	12.	$\log_{27} 3\sqrt{3}$
13.	$\log_{2\sqrt{2}} 8$	14.	$\log_3 27\sqrt{3}$

Sometimes you have to evaluate an expression that has a negative value

Example 6

Evaluate: $\log_3 \frac{1}{9}$

Solutions:

$$\begin{aligned} \text{Let } \log_3 \frac{1}{9} &= x \\ 3^x &= \frac{1}{9} \\ 3^x &= 3^{-2} \\ x &= -2 \end{aligned}$$

$\log_3 \frac{1}{9} = -2$

Exercise: Evaluate the following

1.	$\log_2 \frac{1}{4}$	2.	$\log_4 \frac{1}{16}$
3.	$\log_5 \frac{1}{25}$	4.	$\log_9 \frac{1}{9}$
5.	$\log_{10} \frac{1}{100}$	6.	$\log_2 \frac{1}{32}$
7.	$\log_{10} \frac{1}{1000}$	8.	$\log_8 \frac{1}{64}$
9.	$\log_3 \frac{1}{3\sqrt{3}}$	10.	$\log_9 \frac{1}{27\sqrt{3}}$
11.	$\log_8 \frac{1}{16\sqrt{2}}$	12.	$\log_{10} \frac{1}{10\sqrt{10}}$
13.	$\log_5 \frac{1}{25\sqrt{5}}$	14.	$\log_{2\sqrt{2}} \frac{1}{16}$

Sometimes you have to use the definition of logarithms to solve for x.

Example 7

Solve for x: $\log_2 \frac{1}{4} = x$

Solutions:

$$\begin{aligned} \log_2 \frac{1}{4} &= x \\ 2^x &= \frac{1}{4} \\ 2^x &= 2^{-2} \\ x &= -2 \end{aligned}$$

Example 8

Solve for x: $\log_2 x = -3$

Solutions:

$$\begin{aligned} \log_2 x &= -3 \\ 2^{-3} &= x \\ x &= \frac{1}{8} \end{aligned}$$

Exercise: Solve the following

1.	$\log_2 16 = x$	2.	$\log_5 125 = x$
3.	$\log_{10} 10000 = x$	4.	$\log_2 128 = x$
5.	$\log_2 \sqrt{2} = x$	6.	$\log_2 1 = x$
7.	$\log_5 5 = x$	8.	$\log_2 0.5 = x$
9.	$\log_{10} \sqrt{10} = x$	10.	$\log_{25} x = -1$
11.	$\log_{25} x = \frac{1}{2}$	12.	$\log_{81} x = 1$
13.	$\log_{10} x = 0$	14.	$\log_{12} x = 1$

Sometimes you will have to solve more difficult equations.

Example 9

Solve for x: $\log_8 \sqrt{2} = x$

Solutions:

$$\begin{aligned} \log_8 \sqrt{2} &= x \\ 8^x &= \sqrt{2} \\ 2^{3x} &= 2^{1/2} \\ 3x &= \frac{1}{2} \\ x &= \frac{1}{6} \end{aligned}$$

Example 10

Solve for x: $\log_{16} x = \frac{-3}{4}$

Solutions:

$$\begin{aligned} \log_{16} x &= \frac{-3}{4} \\ 16^{-3/4} &= x \\ x &= (2^4)^{-3/4} \\ x &= \frac{1}{8} \end{aligned}$$

Exercise: Solve the following

1.	$\log_4 \sqrt{2} = x$	2.	$\log_9 \sqrt{3} = x$
3.	$\log_5 \frac{1}{25} = x$	4.	$\log_{49} \sqrt{7} = x$
5.	$\log_{100} \sqrt{10} = x$	6.	$\log_{64} \sqrt{2} = x$
7.	$\log_9 \sqrt{27} = x$	8.	$\log_8 \frac{1}{4} = x$
9.	$\log_8 4\sqrt{2} = x$	10.	$\log_{25} 5\sqrt{5} = x$
11.	$\log_{100} x = -\frac{1}{2}$	12.	$\log_{100} x = \frac{3}{2}$
13.	$\log_{100} x = -\frac{3}{2}$	14.	$\log_8 x = -\frac{1}{3}$
15.	$\log_8 x = \frac{2}{3}$	16.	$\log_{32} x = -\frac{2}{5}$
17.	$\log_5 x = -2$	18.	$\log_7 x = -1$
19.	$\log_{64} x = -\frac{1}{6}$	20.	$\log_{64} x = -\frac{7}{6}$

Logarithms in base 10

Note: when no base is indicated, we assume it means base 10.

Example 11

Evaluate: $\log(10\sqrt{10})$

Solutions:

$$\begin{aligned} \log(10\sqrt{10}) &= \log_{10}(10\sqrt{10}) \\ \text{let } \log_{10}(10\sqrt{10}) &= x \\ 10^x &= 10\sqrt{10} \\ 10^x &= 10^{3/2} \\ x &= \frac{3}{2} \end{aligned}$$

$\therefore \log 10\sqrt{10} = \frac{3}{2}$

Exercise: Solve the following

1.	$\log\sqrt{10}$	2.	$\log 10000$
3.	$\log 1$	4.	$\log\sqrt[3]{10}$
5.	$\log 1000\sqrt{10}$	6.	$\log 0.01$

Exercise: Use your calculator to evaluate correct to 3 decimal places.

1.	$\log_{10} 23$	2.	$\log_{10} 94$
3.	$\log_{10} 7$	4.	$\log_{10} 0.02$
5.	$\log_{10} 0.48$	6.	$\log_{10} 925$

Exercise: Without the use of your calculator, estimate the following by considering powers of 10 and find the integer vales of a and b

7.	$a < \log_{10} 89 < b$	8.	$a < \log_{10} 9 < b$
9.	$a < \log_{10} 0.9 < b$	10.	$a < \log_{10} 498 < b$

Exercise: Now, with different bases, find the integer values of x and y that satisfy the following

11.	$x < \log_{3} 29 < y$	12.	$x < \log_{5} 120 < y$
13.	$x < \log_{2} 33 < y$	14.	$x < \log_{4} 65 < y$

Solving Log Equations

Example 12

Solve: $\log_x 2 = -1$

Solutions:

$$x^{-1} = 2$$
$$\frac{1}{x} = 2$$
$$1 = 2x$$
$$x = \frac{1}{2}$$

or

$$x^{-1} = 2$$
$$(x^{-1})^{-1} = (2)^{-1}$$
$$x = \frac{1}{2}$$

Example 13

Solve: $\log_x 27 = \frac{3}{2}$

Solutions:

$$x^{3/2} = 27$$
$$(x^{3/2})^{\frac{1}{3}} = (27)^{\frac{1}{3}}$$
$$x^{1/2} = 3$$
$$x = 9$$

Exercise: Solve the following

1.	$\log_x 2 = 1$	2.	$\log_x 4 = 2$
3.	$\log_x 9 = 2$	4.	$\log_x 100 = 2$
5.	$\log_x 5 = -1$	6.	$\log_x \sqrt{5} = \frac{1}{2}$
7.	$\log_x 32 = 5$	8.	$\log_x 27 = 3$
9.	$\log_x \frac{1}{8} = -1$	10.	$\log_x 5 = 1$
11.	$\log_x 100 = \frac{1}{2}$	12.	$\log_x 4 = \frac{1}{2}$
13.	$\log_x 8 = \frac{3}{2}$	14.	$\log_x 81 = 4$

Discover some properties of logs

While evaluating the following, try to discover some laws for logarithms

Exercise: Solve the following

1.	$\log_2 2$	2.	$\log_5 5$
3.	$\log_{10} 10$	4.	$\log_8 8$
5.	$\log_{13} 13$	6.	$\log_6 6$
7.	$\log_m m$	8.	$\log_a a$
9.	$\log_2 1$	10.	$\log_5 1$
11.	$\log_{10} 1$	12.	$\log_9 1$
13.	$\log_{12} 1$	14.	$\log_a 1$
15.	$\log_c 1$	16.	$\log_n 1$

Complete the properties and justify using index form.

- $\log_a a =$
- $\log_a 1 =$

On the previous page, you probably noticed:

$\log_a a = 1$	since $a^1 = a$

$\log_a 1 = 0$	since $a^0 = 1$

Now you will need to learn a new law.

Law 1

$$\log_a m + \log_a n = \log_a mn$$

$$\log_a mn = \log_a m + \log_a n$$

Example 14

Simplify: $\log_{10} 2 + \log_{10} 5$

Solutions:

$$\begin{aligned} \log_{10} 2 + \log_{10} 5 &= \log_{10} 2 \times 5 \quad \text{using Law 1} \\ &= \log_{10} 10 \\ &= 1 \end{aligned}$$

Exercise: Evaluate the following

1.	$\log_6 2 + \log_6 3$	2.	$\log_{12} 3 + \log_{12} 4$
3.	$\log_{100} 25 + \log_{100} 4$	4.	$\log_5 20 + \log_5 \frac{1}{4}$
5.	$\log_{10} \sqrt{10} + \log_{10} \sqrt{10}$	6.	$\log_{10} \sqrt{10} + \log_{10} \frac{1}{\sqrt{10}}$
7.	$\log_{15} 5 + \log_{15} 3$	8.	$\log_{20} 5 + \log_{20} 4$
9.	$\log_{10} 4 + \log_{10} \frac{1}{4}$	10.	$\log_{10} 10 + \log_{10} \frac{1}{10}$
11.	$\log_a \sqrt{a} + \log_a \sqrt{a}$	12.	$\log_a a + \log_a \frac{1}{a}$
13.	$\log_a \frac{m}{n} + \log_a \frac{n}{m}$	14.	$\log_a \frac{xy}{z} + \log_a \frac{z}{xy}$

Law 2

$$\log_a m - \log_a n = \log_a \tfrac{m}{n}$$
$$\log_a \tfrac{m}{n} = \log_a m - \log_a n$$

Example 15

Simplify and evaluate: $\log_3 12 - \log_3 4$

Solutions:

$$\begin{aligned} \log_3 12 - \log_3 4 &= \log_3 \frac{12}{4} \quad \text{using Law 2} \\ &= \log_3 3 \\ &= 1 \end{aligned}$$

Exercise: Simplify and evaluate the following

1.	$\log_5 10 - \log_5 2$	2.	$\log_3 6 - \log_3 2$
3.	$\log_5 20 - \log_5 4$	4.	$\log_7 21 - \log_7 3$
5.	$\log_3 24 - \log_3 8$	6.	$\log_8 40 - \log_8 5$
7.	$\log_{10} 30 - \log_{10} 3$	8.	$\log_5 100 - \log_5 20$
9.	$\log_a 3a - \log_a 3$	10.	$\log_c 10c - \log_c 10$
11.	$\log_a a^3 - \log_a a^2$	12.	$\log_c c^5 - \log_c c^4$
13.	$\log_a \sqrt{a} - \log_a \sqrt{a}$	14.	$\log_a \frac{m}{n} - \log_a \frac{m}{n}$

Law 3

$$\log_a m^p = p\log_a m$$

$$p\log_a m = \log_a m^p$$

It follows:

$$\log_a a^x = x$$

We can justify the statement:

$$\begin{aligned} \text{Since } \log_a a^x &= x\log_a a \\ &= x \qquad (\text{as } \log_a a = 1) \end{aligned}$$

Example 16

Simplify: $\log_2 2^4$

Solutions:

$$\log_2 2^4 = 4$$

Example 17

Simplify: $\log_3 81$

Solutions:

$$\begin{aligned} \log_3 81 &= \log_3 3^4 \\ &= 4 \end{aligned}$$

Exercise: Simplify and evaluate the following

1.	$\log_5 125$	2.	$\log_{10} 1000$
3.	$\log_4 64$	4.	$\log_3 \sqrt{3}$
5.	$\log_4 \frac{1}{4}$	6.	$\log_5 \frac{1}{\sqrt{5}}$
7.	$\log_6 36\sqrt{6}$	8.	$\log_4 \frac{1}{16}$
9.	$\log_5 \sqrt{125}$	10.	$\log_2 \frac{1}{32}$

Further Applications of Laws

Example 18

Simplify: $\log_{10} 25 + \log_{10} 40$

Solutions:

$$\begin{aligned} \log_{10} 25 + \log_{10} 40 &= \log_{10} 1000 \\ &= \log_{10} 10^3 \\ &= 3 \end{aligned}$$

Example 19

Simplify: $\log_7 7\sqrt{3}$

Solutions:

$$\begin{aligned} \log_7 7\sqrt{3} &= \log_7 7 + \log_7 \sqrt{3} \\ &= 1 + \log_7 \sqrt{3} \\ &= 1 + \frac{1}{2}\log_7 3 \end{aligned}$$

Exercise: Simplify and the following leaving your answer in exact form

1.	$\log_3 36 - \log_3 4$	2.	$\log_2 32 - \log_2 4$
3.	$\log_{10} 300 - \log_{10} 3$	4.	$\log_{10} 2500 + \log_{10} 4$
5.	$\log_{10} 5 + \log_{10} 4 - \log_{10} 2$	6.	$\log_2 4\sqrt{5}$
7.	$\log_3 81\sqrt{2}$	8.	$\log_3 \frac{9}{5}$
9.	$\log_2 \frac{7}{8}$	10.	$\log_5 \frac{25}{9}$

Sometimes you are asked to expand an expression using the log laws.

Example 20

Expand and simplify: $\log x^2y^3$

Solutions:

$$\begin{aligned} \log x^2y^3 &= \log x^2 + \log y^3 && (\textit{Law } 1) \\ &= 2\log x + 3\log y && (\textit{Law } 3) \end{aligned}$$

Exercise: Simplify and the following leaving your answer in exact form

1.	$\log xy$	2.	$\log \sqrt{xy}$
3.	$\log \frac{x}{y}$	4.	$\log \sqrt{\frac{x}{y}}$
5.	$\log \frac{x+y}{x-y}$	6.	$\log (x+y)^2$
7.	$\log \frac{1}{\sqrt{(x+y)}}$	8.	$\log \sqrt{\frac{x+y}{x-y}}$
9.	$\log \sqrt{x^2+y^2}$	10.	$\log x^2\sqrt{y}$
11.	$\log x^2yz^3$	12.	$\log \frac{x}{y+z}$
13.	$\log \frac{1}{xyz}$	14.	$\log \sqrt{x}(y+z)$
15.	$\log \frac{x+y}{\sqrt{(y+z)}}$	16.	$\log(x^2+y)$

Log Law Summary

	Laws	Examples
1.	$\log_a a = 1$	$\log_5 5 = 1$
2.	$\log_a 1 = 0$	$\log_5 1 = 0$
3.	$\log_a a^x = x$	$\log_5 5^3 = 3$
4.	$\log_a \frac{1}{a} = -1$	$\log_5 \frac{1}{5} = -1$
5.	$\log_a \sqrt{a} = \frac{1}{2}$	$\log_5 \sqrt{5} = \frac{1}{2}$
6.	$\log_a m + \log_a n = \log_a mn$	$\log_2 5 + \log_2 3 = \log_2 15$
7.	$\log_a m - \log_a n = \log_a \frac{m}{n}$	$\log_2 5 - \log_2 3 = \log_2 \frac{5}{3}$
8.	$\log_a m^p = p \log_a m$	$\log_3 5^2 = 2 \log_3 5$
9.	$\log_a \frac{1}{x} = -\log_a x$	$\log_2 \frac{1}{5} = -\log_2 5$
10.	$\log_a \sqrt[n]{x} = \frac{1}{n} \log_a x$	$\log_2 \sqrt[3]{5} = \frac{1}{3} \log_2 5$

True or False?

State True or False for each of the following. If the base is not given, assume it is 10.

		T/F?			T/F?
1.	$\log_5 5 = 1$		12.	$\dfrac{\log 10}{\log 5} = \log 2$	
2.	$\log_5 1 = 0$		13.	$\dfrac{\log 10}{\log 2} = 5$	
3.	$\log_1 5 = 0$		14.	$\log 10 - \log 5 = \log 2$	
4.	$\log_8 8 = 1$		15.	$\log 5 + \log 2 = 10$	
5.	$\log_8 \frac{1}{8} = -1$		16.	$\log x - \log y = \dfrac{\log x}{\log y}$	
6.	$\log_3 9 = 3$		17.	$\log x - \log y = \log \frac{x}{y}$	
7.	$\log_9 \frac{1}{3} = -\frac{1}{2}$		18.	$\log \frac{x}{y} = \log(x - y)$	
8.	$\log_5 \frac{1}{25} = -\frac{1}{2}$		19.	$\log x^5 \div \log x = 5$	
9.	$\frac{1}{2} \log_2 16 = \log_2 8$		20.	$\log x^5 - \log x = 4 \log x$	
10.	$\log_4 16 = \frac{1}{2}$		21.	$\log_x \sqrt{x} = \frac{1}{2}$	
11.	$\log_{16} 4 = \frac{1}{2}$		22.	$\log_y y^2 = y$	

Sometimes you must use log laws to expand a log expression.

Write each expression using log laws in terms of log2, log3 and log10

Note: It is assumed in the questions that the base is 10

Example 21

Simplify: $\log\sqrt{20}$

Solutions:

$$\begin{aligned}\log\sqrt{20} &= \frac{1}{2}(\log 10 + \log 2)\\ &= \frac{1}{2}(1 + \log 2)\end{aligned}$$

Example 22

Simplify: $\log 5.4$

Solutions:

$$\begin{aligned}\log\frac{54}{10} &= \log 27 + \log 2 - \log 10\\ &= 3\log 3 + \log 2 - 1\end{aligned}$$

Exercise: Write each expression in terms of log2, log3 or a constant number

1.	$\log 6$	2.	$\log 20$
3.	$\log 90$	4.	$\log 2000$
5.	$\log 5$	6.	$\log\frac{1}{2}$
7.	$\log 36$	8.	$\log\sqrt{12}$
9.	$\log 0.2$	10.	$\log 1.5$
11.	$\log 0.64$	12.	$\log 25$
13.	$\log 75$	14.	$\log 0.75$
15.	$\log\frac{8}{27}$	16.	$\log 1.8$

Sometimes you have to write an expression when simplifying logs by writing each expression in terms of a and b and/or a constant number.

For this page let $\log_7 8 = a$ and $\log_7 9 = b$

Example 23

Simplify $\log_7 \frac{21}{2}$ by writing in terms of a and b:

Solutions:

Look to use the numbers 8, 9 and the base 7 to make an expression equal to $\frac{21}{2}$

$$\begin{aligned} \log_7 7 + \log_7 3 - \log_7 2 &= 1 + \frac{1}{2}\log_7 9 - \frac{1}{3}\log_7 8 \\ &= 1 + \frac{1}{2}b - \frac{1}{3}a \end{aligned}$$

Exercise: Simplify the following and write each expression in terms of $\log_7 8 = a$ and $\log_7 9 = b$ or a constant number

1.	$\log_7 2$	2.	$\log_7 \frac{8}{9}$
3.	$\log_7 \frac{7}{8}$	4.	$\log_7 \frac{1}{64}$
5.	$\log_7 72$	6.	$\log_7 56$
7.	$\log_7 63$	8.	$\log_7 \frac{1}{81}$
9.	$\log_7 6$	10.	$\log_7 24$

Quick Recap

If $\log_y x = 1.2$ use log laws to simplify and hence evaluate the following		Find the value of x in the following	
1.	$\log_y x^2$	10.	$\log_3 x = 0$
2.	$\log_y \frac{1}{x}$	11.	$\log_3 x = 1$
3.	$\log_y \sqrt{x}$	12.	$x \log_a 5 = \log_a \frac{1}{5}$
4.	$\log_y xy$	13.	$x \log_5 25 = -4$
5.	$\log_y xy^2$	14.	$-5 \log_5 1 = x$
6.	$\log_y \frac{y}{x}$	15.	$\log_2 \sqrt{32} = x$
7.	$\log_y \frac{\sqrt{y}}{x}$		
8.	$\log_y xy - 2\log_y x$		
9.	$\log_y xy \div \log_y \frac{x}{y}$		

Adding and Subtracting Logs

Example 24

Simplify the following and write each expression by writing as a single logarithm (or constant): $\log 12 - 3\log 6 + \log 18$

Solutions:

$$\begin{aligned}\log 12 - 3\log 6 + \log 18 &= \log 12 - \log 6^3 + \log 18\\ &= \log \frac{12 \times 18}{6 \times 6 \times 6}\\ &= \log 1\\ &= 0\end{aligned}$$

Note:

If the base is not written in, it is understood to be 10.

You can only use the laws

$$\log A + \log B = \log AB \quad \text{and} \quad \log A - \log B = \log \tfrac{A}{B}$$

So, if an expression has $c\log A$, you must first use $c\log A = \log A^c$

Exercise: Simplify the following and write each expression by writing as a single logarithm.

1.	$2\log 3 - \log 2 + \log 6$	2.	$\frac{1}{2}\log 9 - \log 6$
3.	$\log 27 - 2\log 3$	4.	$\log 5 + \log 4 - \log 3$
5.	$\frac{1}{2}\log 49 - 2\log 5$	6.	$2\log 4 + \log 8 - \frac{1}{2}\log 4$
7.	$\frac{1}{2}\log 81 - 3\log 3$	8.	$\log x + \log y + \log z$
9.	$3\log x - 2\log z$	10.	$\log xy + \log yz - \log xyz$
11.	$3\log a - \log a^2$	12.	$\log a(a+1) - 2\log a$

Be careful when you are asked to divide logs.

Example 25

Simplify: $\log 64 \div \log 8$

Solutions:

$$\frac{\log 64}{\log 8} = \frac{2\log 8}{\log 8}$$
$$= 2$$

Example 26

Simplify: $\dfrac{\log_5 32}{\log_5 4}$

Solutions:

$$\frac{\log_5 32}{\log_5 4} = \frac{\log_5 2^5}{2\log_5 2}$$
$$= \frac{5\log_5 2}{2\log_5 2}$$
$$= \frac{5}{2}$$

Exercise: Evaluate the following

1.	$\log_2 125 \div \log_2 5$	2.	$\log_5 81 \div \log_5 27$
3.	$\dfrac{\log_7 128}{\log_7 8}$	4.	$\dfrac{\log_5 27}{\log_5 3}$
5.	$\dfrac{\log_3 27}{\log_3 3}$	6.	$\dfrac{\log_a 27}{\log_a 3}$
7.	$\dfrac{\log_3 625}{\log_3 5}$	8.	$\log_2 10000 \div \log_2 100$
9.	$\log_{10} 32 \div \log_{10} 128$	10.	$\dfrac{\log_9 27 - \log_9 3}{\log_3 81}$
11.	$\dfrac{\log_5 24 - \log_5 3}{\log_5 4}$	12.	$\dfrac{\frac{2}{3}\log_{10} 64}{\log_{10} 32 - \log_{10} 4}$
13.	$\dfrac{\log_{10} 25}{\frac{1}{3}\log_{10} 125}$	14.	$\dfrac{\log a}{\log \sqrt{a}}$

Solving Equations with Logs

If the logarithms have the same base, then:

If $\log_c A = \log_c B$ *then* $A = B$

Example 27

Solve: $\log_a x = \log_a 12$

Solutions:

If $\log_a x = \log_a 12$

then $x = 12$

Example 28

Solve: $\frac{1}{2}\log_5 25 = -\log_5 x$

Solutions:

$$\frac{1}{2}\log_5 25 = -\log_5 x$$
$$\log_5 25^{1/2} = \log_5 x^{-1}$$
$$5 = \frac{1}{x}$$
$$x = \frac{1}{5}$$

Exercise: Solve the following

1.	$\log_{10} 2x = \log_{10}(x+1)$	2.	$\log_{10}(3x-1) = \log_{10}(7-x)$
3.	$\log_5(2x+1) = \log_5(x+7)$	4.	$\log_3 x = \log_3(2x-8)$
5.	$\log_3 8 = \log_3(3x-1)$	6.	$\frac{1}{2}\log_2 4 = -\log_2 x$
7.	$\frac{1}{3}\log_{10} 8 = -\log_{10}(x-1)$	8.	$\log_{10}\sqrt{x} = \frac{1}{2}\log_{10} 16$

Sometimes you will have to solve more difficult logarithmic equations

Example 29

Solve $\log_{10} x = \log_{10} 2 + 1$

Solutions:

$$\therefore \log_{10} x = \log_{10} 2 + \log_{10} 10$$
$$\log_{10} x = \log_{10} 20$$
$$x = 20$$

or

$$\log_{10} x - \log_{10} 2 = 1$$
$$\log_{10} \frac{x}{2} = 1$$
$$\frac{x}{2} = 10$$
$$x = 20$$

Exercise: Solve the following

1.	$\log_5 x + \log_5 4 = \log_5 20$	2.	$\log_5 x - \log_5 2 = \log_5 4$
3.	$\log_{10} x = 3\log_{10} 2 + \log_{10} 4$	4.	$\log_5 x - \log_5(x-2) = \log_5 2$
5.	$\log_4 3 + \log_4(x+1) = \log_4 6$	6.	$\log_6 3x - \log_6(3-2x) = 1$
7.	$\log_3 3 + \log_3(x+1) = 2$	8.	$\frac{1}{2}\log_{10} x = 3\log_{10} 2 - \frac{1}{2}\log_{10} 4$
9.	$\log_{10} x = 3\log_{10}(2) + 1$	10.	$\frac{1}{2}\log_5 x = \log_5(3) - 1$
11.	$\log_{10}(x+1) - 2 = \log_{10} 5$	12.	$\log_3(5+2x) - 1 = \log_3 x$

Log equations that simplify to a quadratic equation

Example 30

Solve: $\log_2(x+10) - \log_2(x-2) = \log_2 x$

Solutions:

$$\log_2\left(\frac{x+10}{x-2}\right) = \log_2 x$$

$$\left(\frac{x+10}{x-2}\right) = x$$

$$x + 10 = x^2 - 2x$$

$$x = 5,\ x = -2 \text{ (not a valid solution)}$$

Note:

Always check that your solution is valid ie. *for* $\log x$, $x > 0$

Exercise: Solve the following

1.	$\log_{10}(x-1)+\log_{10}(x+3) = \log_{10} 5$	2.	$\log_2(x-4)+\log_2(x-3) = \log_2 6$
3.	$\log_{10}(x+1) + \log_{10}(x-2) = 1$	4.	$\log_8(x-1) + \log_8(x+1) = 1$
5.	$\log_7(x-4) + \log_7(x+2) = 1$	6.	$\log_{12} x = 1 - \log_{12}(x-1)$
7.	$\log_{12} 2x + \log_{12}(x+1) = 1$	8.	$\log_3 x + \dfrac{3}{\log_3 x} = 4$
9.	$(\log x)^2 - \log x^2 + 1 = 0$	10.	$(\log_3 x)^2 = 4$

Change of base rule

The change of base rule is useful to simplify some expressions and sometimes to evaluate a log when your Calculator only has a base 10 or base e.

Changing from base a to base b

$$\textit{If } \log_a x = \frac{\log_b x}{\log_b a}$$

Only if a and b and x are positive numbers with $a \neq 1$ and $b \neq 1$

Your calculator may be able to calculate these without using the change of base rule.

Example 31

Evaluate correct to 3 decimal places.

Solutions:

Change $\log_3 5$ to base 10 and simplify

$$\log_3 5 = \frac{\log 5}{\log 3}$$
$$= 1.465$$

Example 32

Simplify by using the change of base rule

Solutions:

$$\log_5 4 \times \log_4 25 = \log_5 4 \times \frac{\log_5 25}{\log_5 4}$$
$$= 2\log_5 5$$
$$= 2$$

Use the change of base rule to write the following in base 10 and evaluate correct to 3 significant figures.		By changing the base to the base given in brackets simplify the following	
1.	$\log_2 5$	7.	$\log_9 27$ (base 3)
2.	$\log_3 12$	8.	$\log_4 128$ (base 2)
3.	$\log_{12} 3$	9.	$\log_{0.25} 64$ (base 4)
4.	$\log_3 1000$	10.	$\log_{16} 32$ (base 2)
5.	$\log_5 0.1$	11.	$\log_{25} 125$ (base 5)

More change of base style questions

Exercise: By letting $\log_2 x = b$, simplify the following and express in terms of b

1.	$\log_{\frac{1}{2}} x$	2.	$\log_4 x$
3.	$\log_{\sqrt{2}} x$	4.	$\log_{2x} x$
5.	$\log_{\frac{2}{x}} x$	6.	$\log_{\frac{2}{x}} 2$
7.	$\log_x \frac{4}{x}$	8.	$\log_{\sqrt{x}} 2x$
9.	$\log_{16} x$	10.	$\log_{16} 32x$
11.	$\log_8 2\sqrt{x}$	12.	$\log_8 \frac{2}{\sqrt{x}}$

Simplifying Harder Logs

Exercise: Find the exact value of the following without using a calculator

13.	$\log_{27} 9$	14.	$\log_{16} 128$
15.	$\log_{\frac{1}{5}} 125$	16.	$\log_{\frac{1}{27}} 81$
17.	$\log_{\frac{1}{8}} 16$	18.	$\log_5 27 \times \log_9 25$
19.	$\log_6 9 \times \log_9 5 \times \log_5 6$	20.	$\log_2 3 \times \log_3 5 \times \log_5 8$

21.	Given that $\log_x 3 = \log_y 9$, Show that $y = x^2$	22.	Given that $\log_x 5 = \log_y 125$, Show that $y = x^3$
23.	Show that $\log_x y = \dfrac{1}{\log_y x}$, hence find the value of $\log_{\sqrt{8}} 2$, $\log_{\sqrt{125}} 5$		

Using logarithms to solve exponential equations

Sometimes you have to use the change of base formula to solve equations.

Example 33

Solve: $3^x = 11$

Solutions:

Method 1

Note: 3 is the base

x is the power

11 is the number

$\therefore \log_3 11 = x$

Some calculators can calculate this directly

$$\frac{\log_{10} 11}{\log_{10} 3} = x$$

$\therefore x = 2.183$

Method 2

Take log to the base 10 of both sides

$$\log_{10} 3^x = \log_{10} 11$$
$$x \log_{10} 3 = 11$$
$$x = \frac{\log_{10} 11}{\log_{10} 3}$$
$$x = 2.183$$

Exercise: Solve the following equations correct to 2 decimal places.

1.	$2^x = 5$	2.	$5^x = 7$
3.	$3^x = 2$	4.	$9^x = 8.9$
5.	$4^x = 3.9$	6.	$5^x = 10$
7.	$4^x = 10$	8.	$3^{1+3x} = 25$
9.	$3^{2x} = 10$	10.	$2^{4x-1} = 7$

Sometimes you have to solve more difficult equations

Example 34

Solve: $3^{x+1} = 5^x$

Solutions:

$$\begin{aligned}(x+1)\log_{10} 3 &= x\log_{10} 5 \\ \log_{10} 3 &= x(\log_{10} 5 - \log_{10} 3) \\ x &= 2.151\end{aligned}$$

(correct to 3 decimal places)

OR

$$\begin{aligned}3 \times 3^x &= 5^x \\ 3 &= \left(\frac{5}{3}\right)^x \\ x &= \log_{\frac{5}{3}} 3 = 2.151\end{aligned}$$

Example 35

Solve: $120 = 5 \times 3^n$

Solutions:

$$\begin{aligned}3^n &= 24 \\ n &= \log_3 24 \\ n &= 2.893\end{aligned}$$

Example 36

Solve: $7^{x-1} = 3^{2x}$

Solutions:

$$\begin{aligned}\frac{7^x}{7} &= 9^x \\ \left(\frac{7}{9}\right)^x &= 7 \\ x &= \log_{\frac{7}{9}} 7 \\ x &= -14.572\end{aligned}$$

Exercise: Solve the following equations correct to 2 decimal places.

1.	$2^{x+3} = 5^x$	2.	$5^{x+1} = 11^{x-1}$
3.	$4^{2x-1} = 5^x$	4.	$2^{3x-2} = 5^{x+3}$
5.	$12^{2x} = 7^{x+5}$	6.	$150 = 4 \times 3^n$
7.	$120 = 100 \times 3^{2n}$	8.	$4^{2x} = 9 \times 4^{x+1}$

Example 37

Solve: $\log_2(\log_4 x) = 0$

Solutions:

$$(\log_4 x) = 1$$
$$x = 4$$

Example 38

Solve: $\log_2(\log_4 y) = 1$

Solutions:

$$\log_4 y = 2$$
$$y = 16$$

Exercise: Solve the following equations

1.	$\log_3(\log_5 x) = 0$	2.	$\log_5(\log_3 x) = 1$
3.	$\log_2(\log_2 x) = 2$	4.	$\log_4(\log_4 x) = 2$
5.	$\log_2(\log_3(\log_5 x)) = 0$	6.	$\log_3(\log_2(\log_4 x)) = 0$

Exercise: Simplify the following

7.	$\log_2(\log_2 256)$	8.	$\log_3(\log_3 27)$
9.	$\log_5(\log_2 32)$	10.	$\log_6(\log_3(\log_2 8))$

How do I prove the identity $a^{\log_a x} = x$?

Example 39

Prove the identity $a^{\log_a x} = x$

Solutions:

Proof: Let $a^{\log_a x} = y$

Converting to logarithmic form

$$\log_a y = \log_a x$$
$$y = x$$

Hence $a^{\log_a x} = x$ is an identity

Example: Prove $4^{\log_4 5} = 5$

let $4^{\log_4 5} = y$

$$\log_4 y = \log_4 5$$
$$y = 5$$

$$\therefore 4^{\log_4 5} = 5$$

Exercise: Use the above method to simplify the following

1.	$3^{\log_3 4}$	2.	$5^{\log_5 1}$
3.	$4^{\log_4 \frac{1}{5}}$	4.	$6^{\log_6 \sqrt{3}}$
5.	$2^{\log_2 7}$	6.	$2^{\log_2 \frac{1}{9}}$

Review

State True or False for each of the following. If the base is not given, assume it is 10.

		T/F?			T/F?
1.	$\log_3 \frac{1}{3} = -1$		14.	$\log_2 \frac{1}{4} = -\frac{1}{2}$	
2.	$\log_3 1 = 0$		15.	$\log_5 25 = 5$	
3.	$\log_4 3 = 1$		16.	$\log_x 12 \times \log_x 4 = \log_x 48$	
4.	$\log p + \log q = \log(p + q)$		17.	$\log_x 12 + \log_x 4 = \log_x 48$	
5.	$\log(p - q) = \log p - \log q$		18.	$\log_x 12 + \log_x 4 = \log_x 16$	
6.	$\dfrac{\log p}{\log q} = \log p - \log q$		19.	$\log_2 6 \times \log_2 6 = \log_2 6^2$	
7.	$\log \frac{p}{q} = \log p - \log q$		20.	$\log_5 100 = \log_5 4 + 2$	
8.	$\log_q p = \dfrac{\log_a p}{\log_a q}$		21.	$\log_3 12 = 1 + \log_3 4$	
9.	$\log_6 2 \times \log_6 3 = 1$		22.	$\log_6 3 + \log_6 2 = 1$	
10.	$\log a \times \log b = \log ab$		23.	$2^{\log_2 5} = 5$	
11.	$\log_b a = \log a - \log b$		24.	$3^{\log_3 1} = 0$	
12.	$\log \dfrac{x}{x-1} = \log x - \log(x - 1)$		25.	$4^{\log_4 4} = 1$	
13.	$\log_{25} 5 = \frac{1}{2}$		26.	$\log_a 32 \div \log_a 4 = \log_a 8$	

Understanding Logarithmic Scales in the Real World

A logarithmic scale is a nonlinear scale used to represent data that covers a large range of values.

Instead of increasing by equal additive amounts (like 1, 2, 3...), it increases by multiplicative factors (like 10, 100, 1000...)

Log scales help make sense of data that spans several orders of magnitude. Instead of being overwhelmed by large numbers, the data becomes easier to compare and analyse.

Example 40

We use the Richter scale to compare earthquakes.each increase of 1 unit increases the magnitude of the earthquake by multiplicative of 10.
A moderate earthquake has magnitude 6 and a smaller one has magnitude 4. How many times larger is the amplitude of the magnitude 6 earthquake compared with the magnitude 4 earthquake?

Solutions:

Difference in magnitude $= 6 - 4 = 2$
Each step multiplies amplitude by 10, so the magnitude 6 quake has 100 times the amplitude of the magnitude 4 quake.

Example 41

Stomach acid has pH 1.5 and milk has pH 6.5. How many times more acidic is stomach acid than milk?
Note: 7 is neutral, so a pH of lower than the number is more acidic. 1pH change is x10 change in acidity concentration.

Solutions:

Difference in pH $= 6.5 - 1.5 = 5$
Factor $=$ x$10^5 = 100000$
Stomach acid is 100,000 times more acidic than milk in terms of hydrogen ion.

Exercise: Use the above method or otherwise, to simplify the following

1.	A minor tremor measures 3.2 and a stronger quake measures 5.7. How many times larger is the stronger quake's amplitude?	3a	The decibel (dB) scale for loudness is logarithmic: each 10dB is x10 intensity. A library reading room is approximately 20 dB and a construction site is approximately 80 dB. How many times more intense is the sound at the construction site?
2.	A vinegar sample has pH 1 and a lemon juice sample has pH 4. How many times more acidic is the vinegar compared with the lemon juice?	b	If sound A is 1000 times more intense than sound B, how many decibels higher is A than B?

Application of Logarithms in Problems

Logarithms are applied in solving exponential growth and decay equations, particularly when we want to calculate the time it will take something to increase or decrease to a specific amount. The general form of an exponential equation is

$$P = A(1 + r)^t,$$

Where:

P is the the amount after time t

A is the initial amount

r is the rate (as a decimal)

t is the 'number' of time

Example 42

The amount of money Maya has saved in her bank account can be modelled using the equation $A = 2500 \times (1 + r)^t$, where t is the time and r is the interest rate. How long will it take Maya to save \$3000 if interest is 6% p.a. compounding

a) Yearly b) Monthly

Solutions:

a) Yearly

$$\begin{aligned} A &= 2500 \times 1.06^t \\ 3000 &= 2500 \times 1.06^t \\ 1.2 &= 1.06^t \\ t &= \frac{\log 1.2}{\log 1.06} = 3.13 \end{aligned}$$

$\therefore$ It will take Maya 4 years to save to \$3000

b) Monthly

$$\begin{aligned} A &= 2500 \times 1.005^t \\ 3000 &= 2500 \times 1.005^t \\ 1.2 &= 1.005^t \\ t &= 36.56 \text{ months} \end{aligned}$$

$\therefore$ It will take Maya 37 months to save to \$3000

Example 43

In a particular city, studies show that the rat population grows exponentially at a rate of 80% per year. If there are initially 500 rats, how long will it take for the population to reach 10,000?
We can write an equation for the increase in population as $P = A(1+r)^t$

Solutions:

From the question, we know that $A = 500$, $r = 0.8$
So, $P = 500(1.8)^t$
To find t when $P = 10000$

$$\begin{aligned} 10000 &= 500(1.8)^t \\ 20 &= 1.8^t \\ t &= \frac{\log 20}{\log 1.8} \\ t &= 5.1 \end{aligned}$$

$\therefore$ it would take 5.1 years to reach a population of 10000.

Exercise: Solve the following problems

1. The population of an ant colony can be modelled using the equation $P = 800 \times 2^{0.4t}$, where time is in weeks.

   a) Find the time taken for the population to reach 1600.

   b) Find the time taken for the population to reach 4000.

2. Another ant colony nearby was modelled using the equation $P = 1000 \times 3^{kt}$, where k is a positive constant and t is the time in weeks. If two weeks later the population had increased to 2500, show that $k = \frac{1}{2}\log_3 2.5$

3. A rare coin is currently valued at \$2000, and it appreciates at a rate of 8% per annum. Its value can be modelled by the equation $P = 2000(1.08)^t$, where t is the time in years. How long to the nearest year will it take for the coin to double in value?

4. Suppose Yoni invest \$10 000 in an investment earning 6% per year. How long in years would it take his investment to accumulate to \$15,000 if the investment can be modelled by the equation $A = 10000(1.06)^t$, where t is the time in years.

5. A town has a population of 47,000 in the year 2025. The town's population has been increasing at the rate of 1.2% per year. If $P = 47000(1.012)^t$ is a function that models the population as a function of time, how many years will it take the population to reach 50,000 people?

6. The temperature T°C of an ice pack which is taken out of a freezer after a time t minutes is modelled on the function $T = -40(0.95)^t + 25$

   a Find the temperature of the ice pack when it is taken out the freezer.

   b Find how long it takes for the ice pack to reach 0°C to the nearest minute.

   c Find how long it will take for the ice pack to reach 20°C to the nearest minute.

7. On a farm a population of 400 mice will double every week. The population of mice after t weeks is given by $P = 400 \times 2^t$

   a Write an expression for time t in terms of P

   b Hence, find how long it will take for the population of mice to reach 4000.

Solving Exponential and Logarithmic Inequalities

Example 44

Solve for x:

$$0.5^x \geq 4$$

Solutions:

Method 1: Taking logs (any base) of both sides

$$\begin{aligned} 0.5^x &\geq 4 \\ \log_{10} 0.5^x &\geq \log_{10} 4 \\ x \log_{10} 0.5 &\geq \log_{10} 4 \\ x &\leq \frac{\log_{10} 4}{\log_{10} 0.5} \\ x &\leq -2 \end{aligned}$$

or

Method 2: Using log laws

$$\begin{aligned} 0.5^x &\geq 4 \\ x &\leq \log_{0.5} 4 \end{aligned}$$

Note: $\log_{10} 0.5 < 0$ hence the direction of the inequality is reversed.

x can be negative

Example 45

Without the use of a calculator solve the inequality

$$\log_5 x < 1$$

Solutions:

$$x < 5^1$$

$$0 < x < 5$$

Note that $x > 0$

Example 46

Without the use of a calculator solve the inequality

$$\log_3 x > 2$$

Solutions:

$$x > 3^2$$

$$x > 9$$

Note:

If $0 < a < 1$, then $a^x \geq k \rightarrow x \leq \log_a k$

If $a > 1$, then $a^x \geq k \rightarrow x \geq \log_a k$

Exercise: Solve the following inequalities leaving your answer to 2 decimal places where necessary

1.	$3^x > 10$	2.	$2^x < 5$
3.	$0.4^x > 7$	4.	$5^{x-1} \leq 12$
5.	$0.8^x < 6$	6.	$\log_{10} x > 2$
7.	$\log_3 x \geq 2$	8.	$\log_2 x < 3$
9.	$\log_5 x \leq 2$	10.	$\log_2 (x-1) \leq 3$

Fallacies

Example 47

Find the fallacy in the following proof

$$\frac{1}{9} > \frac{1}{27}$$

$$\therefore \left(\frac{1}{3}\right)^2 > \left(\frac{1}{3}\right)^3$$

$$\therefore \quad \log\left(\frac{1}{3}\right)^2 > \log\left(\frac{1}{3}\right)^3$$

$$2\log\left(\frac{1}{3}\right) > 3\log\left(\frac{1}{3}\right) \quad \longleftarrow \text{The fallacy is here, as } \log\left(\frac{1}{3}\right) < 0 \text{ so when we divide}$$

$$\therefore \quad 2 > 3 \qquad \text{by it we need to change the direction of the sign}$$

Exercise:

1.	Find the fallacy in the following proof $\frac{1}{2} > \frac{1}{7}$ $\log\frac{1}{2} > \log\frac{1}{7}$ $\frac{\log\frac{1}{2}}{\log\frac{1}{7}} > 1$ $\frac{-\log 2}{-\log 7} > 1$ $\log 2 > \log 7$ $\therefore 2 > 7$	2.	Find the fallacy in the following proof $\frac{1}{8} > \frac{1}{16}$ $\left(\frac{1}{2}\right)^3 > \left(\frac{1}{2}\right)^4$ $3\log\frac{1}{2} > 4\log\frac{1}{2}$ $\therefore 3 > 4$
3.	Start with the inequality $\frac{1}{5} > \frac{1}{125}$ and write a fallacy to show $1 > 3$		

Simultaneous Equations involving Logs

To solve these equations, your skills in solving simultaneous equations, log laws and laws of indices are required.

Example 48

$$\begin{cases} \log_2 x + \log_2 y = 4 & (1) \\ \log_2 x - 2\log_2 y = 1 & (2) \end{cases}$$

Solutions:

from (1) $\quad xy = 16 \quad (3)$

from (2) $\quad \dfrac{x}{y^2} = 2 \rightarrow x = 2y^2 \quad (4)$

substitute (4) into (3)

$$2y^3 = 16$$

$$y = 2, \quad \therefore x = 8$$

OR

(1) − (2)

$$3\log_2 y = 3$$

$$y = 2$$

$$\log_2(x) + 1 = 4$$

$$x = 8$$

Example 49

The points $(25, 0)$ and $\left(\frac{1}{5}, 3\right)$ both lie on the graph $y = a - \log_b x$. Find the value of a and b

Solutions:

$(25, 0)$ $\qquad \left(\frac{1}{5}, 3\right)$

$0 = a - \log_b 25 \quad (1) \qquad 3 = a - \log_b \frac{1}{5} \quad (2)$

$(2) - (1) \quad 3 = 3\log_b 5$

$$b = 5, \quad a = 2$$

Exercise: Solve the following simultaneous equations

1.	$\begin{cases} \log_2 x - \log_2 y = 3 \\ \log_2(x - 2y) = 1 \end{cases}$	2.	$\begin{cases} \log_2 x - \log_2 y = 2 \\ \log_2(x + y) = 3 \end{cases}$
3.	$\begin{cases} \log_4 y = 3 \\ \log_2 y = 4 + \log_2 x \end{cases}$	4.	$\begin{cases} \log_y x = 2 \\ \log_2(x + 2y) = 3 \end{cases}$
5.	$\begin{cases} \log_3 xy^2 = 0 \\ \log_3 x^2 y = 3 \end{cases}$	6.	The points $(27, 20)$ and $(\frac{1}{9}, -5)$ both lie on the graph of the function $y = a\log_3 bx$. Find the value of a and b
7.	The points $(4, 0)$ and $(64, 2)$ both lie on the graph of $y = a + \log_b x$. Find the value of a and b		

Domain for log functions

$\log(f(x))$ is only defined for $f(x) > 0$

Example 50

Find the domain of the logarithmic function $y = \log(1 - x)$

Solutions:

This function is only defined when

$$1 - x > 0$$

$\therefore$ Domain: $x < 1$, *or* $(-\infty, 1)$

Exercise: Find the domain of the following functions

1.	$y = \log(x - 3)$	2.	$y = \log(2x)$
3.	$y = \log(2x + 6)$	4.	$y = \log(3 - x)$
5.	$y = \log\sqrt{x}$	6.	$y = \log(-x)$
7.	$y = -\log(x)$	8.	$y = \log(x + 5)$
9.	$y = \log(1 + x^2)$	10.	$y = \log(x^2)$

Justifications

$$\log_a a = 1 \qquad \text{since } a^1 = a$$

$$\log_a 1 = 0 \qquad \text{since } a^0 = 1$$

Log Law Proofs

Law 1

$$\log_a m + \log_a n = \log_a mn$$
$$\log_a mn = \log_a m + \log_a n$$

Proof 1

Let $m = a^x$ and $n = a^y$ So, $\log_a m = x$ and $\log_a n = y$

$$mn = a^x a^y = a^{x+y}$$

$\therefore \log_a mn = x + y$

$$= \log_a m + \log_a n$$

Proof 2

We will use $m = a^{\log_a m}$ and $n = a^{\log_a n}$

$$\log_a mn = \log_a \left(a^{\log_a m} \times a^{\log_a n}\right)$$

$$= \log_a \left(a^{\log_a m + \log_a n}\right)$$

$$= \log_a m + \log_a n$$

Law 2

$$\log_a m - \log_a n = \log_a \frac{m}{n}$$

$$\log_a \frac{m}{n} = \log_a m - \log_a n$$

Proof 1

Let $m = a^x$ and $n = a^y$ So, $\log_a m = x$ and $\log_a n = y$

$$\frac{m}{n} = \frac{a^x}{a^y} = a^{x-y}$$

$\therefore \log_a \frac{m}{n} = x - y$

$$= \log_a m - \log_a n$$

Proof 2

We will use $m = a^{\log_a m}$ and $n = a^{\log_a n}$

$$\log_a \frac{m}{n} = \log_a \frac{a^{\log_a m}}{a^{\log_a n}}$$

$$= \log_a a^{\log_a m - \log_a n}$$

$$= \log_a m - \log_a n$$

Law 3

$$\log_a m^p = p\log_a m$$

$$p\log_a m = \log_a m^p$$

Proof 1

Let $\log_a m = x$

So $a^x = m$

$m^p = a^{xp}$

$\log_a m^p = xp$

$= p\log_a m$

Proof 2

We will use $m = a^{\log_a m}$

$$\log_a m^p = \log_a (a^{\log_a m})^p$$

$$= \log_a a^{p\log_a m}$$

$$= p\log_a m$$

Change of Base Rule Proof

Changing from base a to base b

$$\log_a x = \frac{\log_b x}{\log_b a}$$

Proof 1

Let $\dfrac{\log_b x}{\log_b a} = y$

$$\log_b x = y\log_b a$$

$$\log_b x = \log_b a^y$$

$$\therefore \quad x = a^y$$

$$\log_a x = y$$

LHS = RHS

Proof 2

Let $y = \log_a x$

$$a^y = x$$

$$\log_b a^y = \log_b x$$

$$y\log_b a = \log_b x$$

$$y = \frac{\log_b x}{\log_b a}$$

$$\log_a x = \frac{\log_b x}{\log_b a}$$

Review 1 (up to page 22)

1 Evaluate the following

a $\log_{10} 10$

b $\log_7 7$

c $\log_5 25$

d $\log_2 16$

e $\log_4 16$

f $\log_5 \frac{1}{5}$

g $\log_5 \sqrt{5}$

h $\log_7 \frac{1}{49}$

i $\log_3 \frac{1}{\sqrt{3}}$

2 Evaluate the following

a $\log_{10} 25 + \log_{10} 4$

b $\log_{10} 4 + \log_{10} \frac{1}{4}$

c $\log_2 18 - \log_2 9$

d $\log_{10} 300 - \log_{10} 3$

e $\log_2 80 - \log_2 5$

f $\log_3 \sqrt{3} + \log_3 \sqrt{3}$

g $\log_5(10x) - \log_5(2x)$

h $\log_{10}\left(\frac{a}{b}\right) + \log_{10}\left(\frac{b}{a}\right)$

i $\log_4 8 + \log_4 2$

j $\log_y x + \log_y \frac{1}{x}$

k $\log_2 24 - \log_2 3$

l $\log_{10} 125 + \log_{10} 8$

3 Evaluate given that $a > 0,\ a \neq 1$

a $\log_a a$

b $\log_a 1$

c $\log_a \frac{1}{a}$

d $\log_a a^3$

e $\log_a \sqrt{a}$

f $\log_a \frac{1}{\sqrt{a}}$

g $\log_a a\sqrt{a}$

h $\log_a \frac{1}{a^2}$

i $\log_a a^2\sqrt{a}$

j $\log_a \sqrt[3]{a}$

k $\log_{\sqrt{a}} a$

l $\log_{\sqrt{a}} \frac{1}{a}$

4 Solve the following

a $\log_3 x = 2$

b $\log_2 x = 3$

c $\log_5 x = 0$

d $\log_2 x = -3$

e $\log_{25} x = -\frac{1}{2}$

f $\log_{10} x = 1$

g $\log_{27} x = -\frac{2}{3}$

5 Solve the following equations

a $\log_2 64 = x$

b $\log_2 \frac{1}{4} = x$

c $\log_8 1 = x$

d $\log_{10} 0.1 = x$

e $\log_x 81 = 4$

f $\log_2 \sqrt{32} = x$

g $\log_{27} \sqrt{3} = x$

h $\log_9 3\sqrt{3} = x$

i $\log_x 125 = 3$

j $\log_x 4 = -1$

6 If $\log_5 a = x$, simplify

a $\log_5 25a$

b $\log_5 \frac{a^3}{5}$

7 If $\log_2 x = 1.44$ evaluate

a $\log_2 \sqrt{x}$

b $\log_2 4x$

8 Simplify the following

a $\frac{1}{2}\log_3 16 - \log_3 2$

b $2\log_{10} 9 \div \log_{10} 27$

c $3\log_2 x + 2\log_2 y$

d $\frac{\log_5 32}{\log_5 4}$

9 Expand using log laws

a $\log x^3 y$

b $\log \sqrt{\frac{x+y}{x-y}}$

10 For what values of x is $\log_a x$

a Positive

b Zero

c Negative

d Undefined

11 Express in logarithmic form

a $5^3 = 125$

b $16 = 64^{2/3}$

c $\sqrt{a} = a^{1/2}$

d $\frac{1}{a} = a^{-1}$

12 Express in index form

a $\log_2 32 = 5$

b $\log_8 4 = \frac{2}{3}$

c $\log_a 1 = 0$

d $\log_a a = 1$

13 Simplify the following

a $\log_a \frac{1}{m}$

b $\log_a \sqrt{m}$

c $\log_a a^x$

d $a^{\log_a x}$

Review 2 (up to page 28)

1 Evaluate the following

a $\log_3 27$

b $\log_3 \frac{1}{27}$

c $\log_3 \frac{1}{9\sqrt{3}}$

d $\log_{10} \sqrt[3]{10}$

e $\log_{27} 3$

f $\log_{10} 0.001$

2 Solve the following

a $\log_8 32 = x$

b $\log_3 \frac{1}{9} = x$

c $\log_{\sqrt{6}} 36 = x$

d $\log_{32} 16 = x$

e $\log_{16} \frac{1}{8\sqrt{2}} = x$

f $\log_5 x = \log_3 9$

g $\log_x 5 = \frac{1}{2}$

h $\log_2 x = -\frac{1}{2}$

i $\log_x \frac{1}{3} = -1$

j $\log_9 x = 2.5$

k $\log_x 2 = -\frac{1}{2}$

l $\log_x 36 = 2$

3 Solve the following equations

a $\log_{10} 3x = \log_{10}(x+2)$

b $\frac{1}{2}\log_4 16 = \log_6 x$

c $\log 4x - \log 3 = \log(x+4)$

d $2\log_5 x = \log_5 49$

e $\log_{10} x + \log_{10} 5 = \log_{10}(x+1)$

f $\log_3(2x+1) - \log_3(2x-1) = 1$

g $\log_y x = 2\log_y 3 - 3\log_y 2$

h $-3 + \log_5 20 = \log_5 4 - \log_5 x$

4 Evaluate the following

a $\log_5 25 - \log_3 \frac{1}{27}$

b $\log_{36} 6 + \log_{25} 5$

c $\log_{10} 27 \div \log_{10} 81$

d $\log_5 0.2 + \log_5 0.04$

e $\log_2 16 \div \log_2 8$

f $\log_2 40 - \log_2 5$

g $\frac{\log_4 27}{\log_4 \sqrt{3}}$

5 If $\log_a 2 = p$, $\log_a 5 = q$ and $\log_a 12 = r$ Express the following in terms of p, q and/or r.

a $\log_a 10$

b $\log_a 6$

c $\log_a \frac{1}{2}$

d $\log_a 144$

e $\log_a 125 + \log_a 4$

f $\log_a 3$

g $\log_a 3a$

h $\log_a 36$

i $\log_a 5\sqrt{5}$

6 If $\log_{10} x = y$ express each expression in terms of y

a $\log_{10} x^2$

b $\log_{10} \frac{1}{x^3}$

c $\log_{10} \sqrt{x}$

d $\log_{10} 10\sqrt{x}$

e $\log_{10} 100x$

7 If $\log_2 5 = x$, find the value of 8^x

8 Explain why $\log_{10}(0.03) = -2 + \log_{10} 3$

9 Simplify by changing the base to the given value

a $\log_8 128$ (base 2)

b $\log_9 243$ (base 3)

10 Express $1 + \log_5 3$ as a single logarithm

11 If $\log_x 9 = 1.2$ and $\log_x 18 = 1.6$, find the value of

a $\log_x 2$

b $\log_x 162$

c $\log_x 3$

d $\log_x 6$

e $\log_x 27$

f $\log_x 4.5$

12 Explain why $\log_{10} 253$ must lie between 2 and 3

13 Explain why $\log_2 12$ lies between 3 and 4

14 Solve the following equations

a $7^x = 11$

b $4^{x+1} = 7^2$

Review 3

1 If
$\log_3 \frac{1}{9} + \log_3 \frac{1}{3} + \log_3 1 + \log_3 3 + \log_3 9 = k$

Find k

2 Simplify the following

a $\log y + \log xy + \log xy^2 + \log xy^3$

b $\log \frac{1}{x} + \log \frac{1}{x^2} + \log \frac{1}{x^3}$

3 Evaluate the following

a $\log_a 8 \div \log_a 4$

b $[\log_{10}(5 \log_{10} 100)]^3$

c $\log_2(\log_{10}(10000))$

d $\dfrac{\log_3 18}{2 + \log_3 2}$

e $\dfrac{\log_5 \sqrt{75}}{2 + \log_5 3}$

f $\dfrac{\log_3 \sqrt{162}}{4 + \log_3 2}$

4 Evaluate using change of base rule

a $\log_2 3 \times \log_3 6 \times \log_6 8$

b $\log_{10} 64 \times \log_3 10 \div \log_3 4$

c $\dfrac{\log_{16} 4}{\log_9 3}$

5 Prove the following

a $\log_y x = \dfrac{1}{\log_x y}$

b $\log_y x \times \log_x y = 1$

6 Express x in terms of y

a $\log_3 x + \log_3 4 = \log_3 12 - \log_3 y$

b $\log_2 x = 3 + \log_2 y$

c $\log_3 x = 4 - 2\log_3 y$

d $\log_{10} x + 2\log_{10} y = 1$

7 Prove $a^{\log b} = b^{\log a}$
Hint: Let $x = a^{\log b}$ and $y = b^{\log a}$
and take logs of both sides in
each expression.

8 If $p = \log_{10} 2$ and $q = \log_{10} 3$
Express each of the following
in terms of p and q.

a $\log_{10} 0.3$

b $\log_{10} \frac{1}{25}$

c $\log_{10} 6.4$

9a Eliminate 'a' from the equations
$x = \log_a p$
$y = \log_a q$

b Try solving the above question
in different ways

10 Simplify
$\log 6! - \log 5! + \log 4! - \log 3! + \log 2! - \log 1!$
Note: $6! = 6 \times 5 \times 4 \times 3 \times 2 \times 1$
$3! = 3 \times 2 \times 1$

11 Consider the equation $3^x = 30$

a Explain why the solution to this
equation lies

between $x = 3$ and $x = 4$

b Solve this equation giving your answer
corrected to 2 decimal places

12 Solve these equation giving your answer corrected to 2 decimal places

a $5^x + 5^{x+1} = 18$

b $9^{2x} = 5^{x+3}$

c $9^x = 6 \times 5^x$

d $7^{3x} = 2 \times 7^{x+1}$

e $120 = 9 \times 5^x$

13 Solve the following equations

a $x \log_2 5 = \log_{10} 5$

b $2 \log_2 x - \log_2 x^3 = 3$

c $(\log_2 x)^2 = 6 \log_2 x - 8$

d $3(\log_8 x)^2 - \log_8 x^{10} = -8$

e $(\log_3 x)^2 = \log_3 x^5$

14 Joel has \$3000 to invest in an account that pays 4.8% p.a. How long will it take for this investment to grow to \$5000 if investment is compounded:

a Annually

b Quarterly

15 A cup of tea at 95°C is placed in the fridge with the temperature at 0°C. Newton's Law of cooling applies so that the temperature halves every 30 minutes. The temperature T°C after t hours in the fridge is modelled by the function

$$T = 95 \times \left(\frac{1}{2}\right)^{2t}$$

a What is the temperature of the cup of tea after 2 hours to the nearest degree?
Note: $2t$ is the number of 30 minute periods in t hours.

b How long (to the nearest minute) will it take for the temperature to fall to 2°C?

16 Evaluate the following

a $\log_3(\log_4 64)$

b $3^{\log_3 6}$

c $5^{2 \log_5 3}$

17 If $y = \log_3 7$, Find 9^y

18 Given $y = 2 \times 10^{5x}$, express x in terms of y.

19 Solve the following inequalities

a $5^x < 12$

b $0.5^x > 3$

c $\log_2 x < 5$

20 The value of a brand new car is \$38,000. The car depreciates at a rate of 10% per year according to the formula $P = 38\,000(0.9)^t$, where t is the number of years. During which year will the value of the car drop below half its original price?

Extension

1 Simplify $\dfrac{1}{\log_3 a} + \dfrac{1}{\log_4 a} + \dfrac{1}{\log_5 a}$

2 If $\log_{10} 3 = a$ and $\log_3 5 = b$ then express $\log_{10} 5$ in terms of a and b

3 Express x in terms of y

a $x = \log_{27} 169$ and $y = \log_3 13$

b $x = \log_{25} 81$ and $y = \log_5 3$

4 Simplify $\log_2 9 \times \log_5 8 \times \log_3 \dfrac{1}{5}$

5 Given that $x^2 + y^2 = 23xy$

a Express $\left(\dfrac{x+y}{5}\right)^2$ in terms of x and y.

b Hence show that

$$\log\left(\frac{x+y}{5}\right) = \frac{1}{2}(\log x + \log y)$$

6 If $p = \log_{10} 2$ and $q = \log_{10} 3$, Express $\log_5 12$ in terms of p and q

7 If $p = \log_8 225$ and $q = \log_2 15$, Express p in terms of q

8 Find the value of x if

a $\log_2(\log_3(\log_4 x)) = 0$

b $\log_4(\log_2(\log_3 x)) = 0$

9 If $\log 9 = p$ and $\log 18 = r$ Show $\log \dfrac{4}{3} = 2r - \dfrac{5}{2}p$

10 Solve the following (correct to 2 d.p where necessary)

a $4^{2x} = 9 \times 4^{x+1}$

b $\log_{\sqrt{x}} 2 + \log_4 x = 2$

c $\log_{\sqrt{x}} 3 + \log_9 x = 2$

d $\log_2 x + \dfrac{4}{\log_2 x} = 5$

11 Show that

$$\log_{ab} x = \frac{\log_a x}{1 + \log_a b} = \frac{\log_b x}{1 + \log_b a}$$

12 Find the domain of the following log functions

a $y = \log|x|$

b $y = \log(4 - x^2)$

c $y = \log(x^2 - 4)$

d $y = \log|4 - x|$

e $y = \log\sqrt{4 - x^2}$

f $y = \log x + \log(x - 4)$

13 If $\log \dfrac{x^2}{y} = 1$ and $\log xy^3 = 1$ what is the value of

a $\dfrac{\log x}{\log y}$

b $\log \dfrac{x}{y}$

14 Find the line where the fallacy occurs:

$\dfrac{1}{2} > \dfrac{1}{4}$

$\log \dfrac{1}{2} > \log \left(\dfrac{1}{2}\right)^2$

$\log \dfrac{1}{2} > 2 \log \dfrac{1}{2}$

$\therefore 1 > 2$

15 Prove

a $\log_{\frac{1}{a}} x = -\log_a x$

b $\log_{\frac{1}{a}} a = -1,\ a > 0,\ a \neq 1$

Graphs of Logarithmic Functions

We will start by looking at the equation $y = 2^x$.

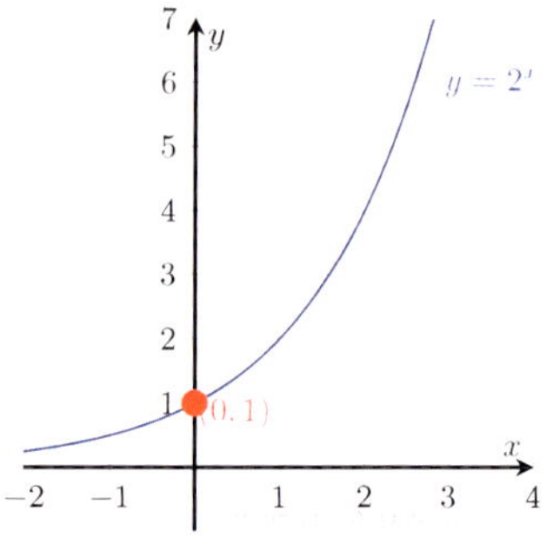

We know that this function has the properties

- Domain: all real numbers $x \in \mathbb{R}$
- Range: $y > 0$
- Cuts the y-axis at $y = 1$
- Increases rapidly as x increases
- Horizontal asymptote $y = 0$

The inverse of $y = 2^x$ is $x = 2^y \Rightarrow y = \log_2 x$.

Graphically, the inverse of a functions is the reflection about the line $y = x$.

So, if we take the graph of $y = 2^x$ and reflect it about the line $y = x$, we get the graph of $y = \log_2 x$

- Domain: all real numbers $x > 0$
- Range: $y \in \mathbb{R}$
- Cuts the x-axis at $x = 1$
- Increases slowly as x increases
- Vertical asymptote $x = 0$

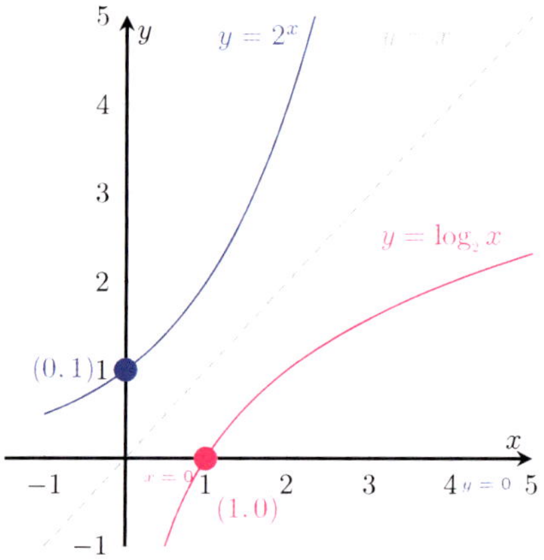

In general, we have similar properties for $y = a^x$ ***and*** $y = \log_a x$, $a > 1$

$y = a^x$

- Domain: all real x, $x \in \mathbb{R}$
- Range: $y > 0$
- Cuts the y-axis at $y = 1$
- Increases quickly as x increases
- Horizontal asymptote $y = 0$

$y = \log_a x$, $a > 1$

- Domain: $x > 0$
- Range: all real y, $y \in \mathbb{R}$
- Cuts the x-axis at $x = 1$
- Increases slowly as x increases
- Vertical asymptote $x = 0$

Exercise:

1. Complete the table of values for $y = \log_2 x$ and plot these points on the graph below

x	$\frac{1}{8}$	$\frac{1}{4}$	$\frac{1}{2}$	1	2	4	8
$y = \log_2 x$			-1	0			

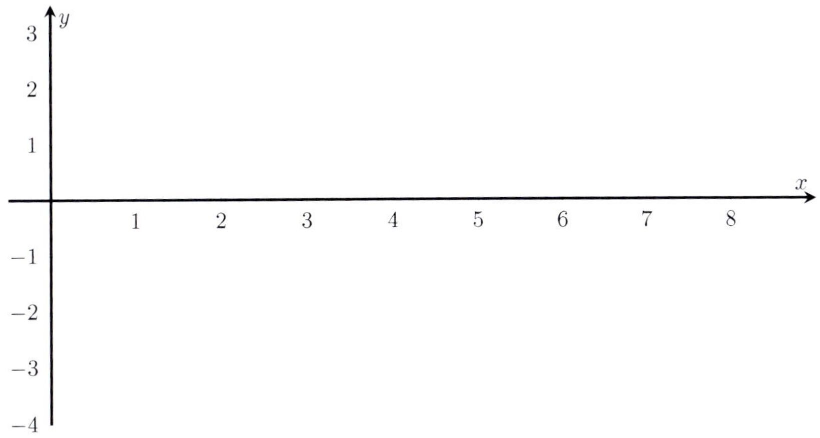

2. Complete the table of values for $y = \log_3 x$ and plot these points on the graph above

x	$\frac{1}{27}$	$\frac{1}{9}$	$\frac{1}{3}$	1	3	9	27
$y = \log_3 x$			-1	0			

3a. By using any appropriate graphics technology draw the graphs of

- $y = \log_2 x$
- $y = \log_3 x$
- $y = \log_{10} x$
- $y = \log_a x$, $0 < a < 1$
- $y = \log_a x$, $1 < a < 10$

b. What happens to the shape of the curve as the base a increases?

c. Does the domain or range change?

d. Does the vertical asymptote change as a varies?

e. Why do all the graphs pass through the point $(1, 0)$?

Vertical Translation

Equation (any positive base)	Transformation from $y = \log_2 x$	Point Transformation
$y = \log_2 x + c$	Vertical translation up c units	$(x, y) \to (x, y + c)$
$y = \log_2 x - c$	Vertical translation down c units	$(x, y) \to (x, y - c)$

Example 51

The graph of $y = \log x$ is translated up 1 unit.

a Write the equation of the new function.

b Sketch both graphs and label the x-intercepts

Solutions:

a $y = \log x + 1$

b

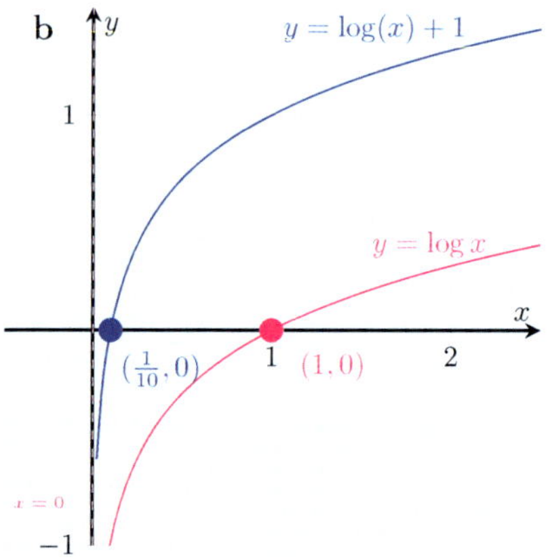

Example 52

The point $(1, 0)$ lies on $y = \log x$.
Find the corresponding point on $y = \log x + 5$
Solutions:
$(1, 0) \to (1, 5)$

Example 53

The point $(100, -2)$ lies on $y = \log(x) - 4$
Find the corresponding point on $y = \log x$
Solutions:
$(100, -2) \to (100, -2 + 4) \to (100, 2)$

Exercise:

1. The graph of $y = \log x$ is translated up 2 units. Write the equation of the new function.
2. The graph of $y = \log x$ is translated down 4 units. Write the equation of the new function.
3. The graph of $y = \log x$ is translated to $y = \log(x) - 3$. Describe the transformation in words.
4. The point $(10, 1)$ lies on $y = \log x$. What is the corresponding point on $y = \log(x) + 3$?
5. The point $\left(\frac{1}{10}, -1\right)$ lies on $y = \log(x)$. Find the corresponding point on $y = \log(x) - 2$.
6. The point $(10, 4)$ lies on $y = \log(x) + 3$. Find the corresponding point on $y = \log x$

Horizontal Translation

Equation (any positive base)	Transformation from $y = \log_2 x$	Point Transformation
$y = \log_2(x + b)$	Horizontal translation left b units	$(x, y) \to (x - b, y)$
$y = \log_2(x - b)$	Horizontal translation right b units	$(x, y) \to (x + b, y)$

Example 54

The graph of $y = \log x$ is translated left 1 unit.

a Write the equation of the new function.

b Sketch both graphs and label the x-intercepts

Solutions:

a $y = \log(x + 1)$

b

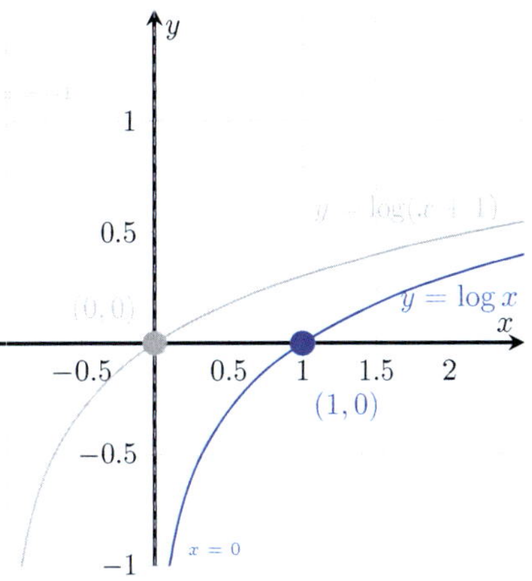

Example 55

The point $(1, 0)$ lies on $y = \log x$.
Find the corresponding point on $y = \log(x - 4)$.

Solutions:

$(1, 0) \to (5, 0)$

Example 56

The point $(98, 2)$ lies on $y = \log(x + 2)$.
Find the corresponding point on $y = \log x$

Solutions:

$(98, 2) \to (100, 2)$

Exercise:

1. The graph of $y = \log x$ is translated right 2 units. Write the equation of the new function.
2. The graph of $y = \log x$ is translated left 4 units. Write the equation of the new function.
3. The graph of $y = \log x$ is translated to $y = \log(x + 2)$. Describe the transformation in words.
4. The point $(10, 1)$ lies on $y = \log x$. What is the corresponding point on $y = \log(x - 5)$?
5. The point $(12, 2)$ lies on $y = \log_3(x - 3)$. What is the original point on $y = \log_3 x$

Horizontal and Vertical Translations

Transformations can be applied in various orders. For consistency, this workbook always applies horizontal transformations before vertical transformations.

Equation (any positive base)	Transformation from $y = \log_2 x$	Point Transformation
$y = \log_2(x+b) + c$	Horizontal translation left b units and Vertical translation up c units	$(x, y) \to (x - b, y + c)$
$y = \log_2(x-b) - c$	Horizontal translation right b units and Vertical translation down c units	$(x, y) \to (x + b, y - c)$

Example 57

The graph of $y = \log x$ is translated left 1 unit then up 2.

a Write the equation of the new function.

b Sketch both graphs and label the y-intercept

Solutions:

a $\quad y = \log(x+1) + 2$

b.

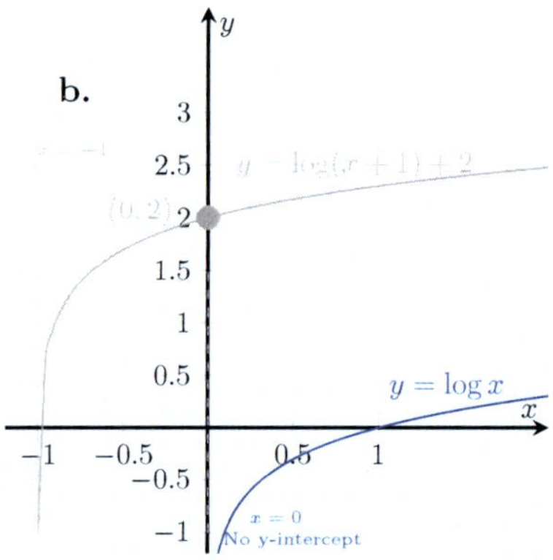

Example 58

The point $(1, 0)$ lies on $y = \log x$. Find the corresponding point on $y = \log(x-4) + 3$.

Solutions:

$(1, 0) \to (5, 3)$

Example 59

The point $(99, 0)$ lies on $y = \log(x+1) - 2$. Find the corresponding point on $y = \log x$

Solutions:

$(99, 0) \to (100, 2)$

Exercise:

1. Write the equation of the new function when the graph of $y = \log x$ is translated right 2 units then down 1 unit.
2. Write the equation of the new function after the graph of $y = \log x$ is translated left 4 units then down 3 unit.
3. The graph of $y = \log x$ is translated to $y = \log(x+2) - 3$. Describe the transformation in words.
4. The point $(100, 2)$ lies on $y = \log x$. What is the corresponding point on $y = \log(x+4) - 2$?
5. The point $(8, 3)$ lies on $y = \log_5(x-3) + 2$. What is the original point on $y = \log_5 x$?

Reflection about the X and Y axes

Equation (any positive base)	Transformation from $y = \log_2 x$	Point Transformation
$y = -\log_2 x$	Reflect about the x-axis	$(x, y) \to (x, -y)$
$y = \log_2(-x)$	Reflect about the y-axis	$(x, y) \to (-x, y)$

Example 60

Draw the graphs of $y = \log_2 x$ and $y = -\log_2 x$

Solutions:

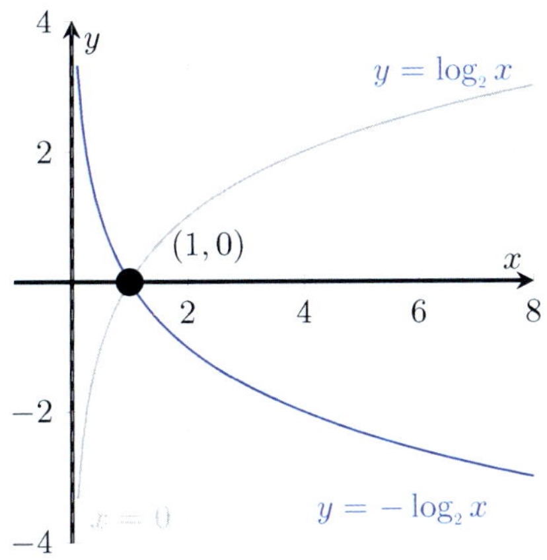

Example 61

Draw the graphs of $y = \log_4 x$ and $y = \log_4(-x)$

Solutions:

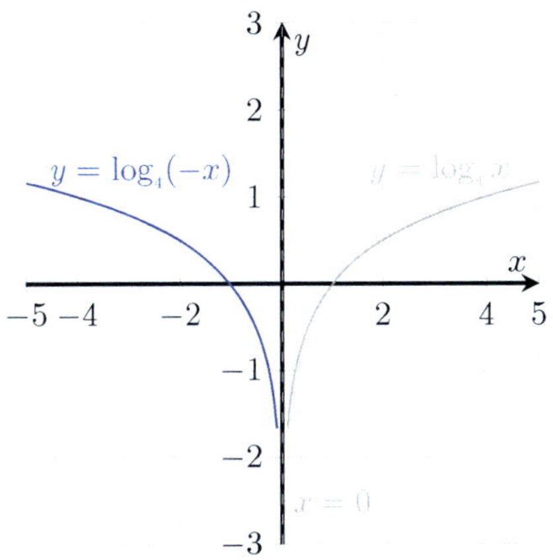

Exercise:

1. The point $(9, 2)$ lies on the graph of $y = \log_3 x$. Find the point when the graph is reflected about:

   a the y-axis
   b the x-axis
   c Both the x and y-axes

2. On the same set of axes draw $y = \log_3 x$ and $y = \log_3(-x)$ and $y = -\log_3 x$

3. If the point $(-64, 3)$ lies on the function $y = \log_4(-x)$, what is the corresponding point that lies on the function $y = -\log_4 x$

Vertical Dilation

Equation (any positive base)	**Transformation from $y = \log_2 x$**	**Point Transformation**
$y = k\log_2 x, k > 0$	Vertical dilation of scale factor k	$(x, y) \to (x, ky)$

Note: We say there is a dilation in the y direction by a scale factor of k

or

We can also say there is a dilation from the x-axis by a scale factor of k

When $k > 1$, we say there is a vertical stretch of k

When $0 < k < 1$, we say there is a vertical compression of k

Example 62

If the point $(9, 4)$ lies on the graph of $y = 2\log_3(x)$, what point lies on the parent function $y = \log_3(x)$ and describe the transformation in words

Solutions:

$(x, 2y) \to (x, y)$

So, $2y = 4$

$(9, 4) \to (9, 2)$

or

$y = 2\log_3 x$

$\frac{y}{2} = \log_3 x$

So, $y = 4 \to 2$

Hence $(x, y) = (9, 2)$

There is a vertical compression of scale factor $\frac{1}{2}$

Exercise:

1. Write the new equation obtained when the graph $y = \log x$ is dilated vertically by a scale factor of 2 units.

2. The graph of $y = \log_3 x$ is dilated to $y = 4\log_3 x$

   a Describe the transformation

   b Find the point where $y = 4\log_3 x$ crosses the x-axis

3. The point $(16, 12)$ lies on $y = 3\log_2 x$. What is the corresponding point on $y = 5\log_2 x$?

4. The point $\left(\frac{1}{4}, -2\right)$ lies on $y = \log_2 x$. What is the corresponding point on $y = \frac{1}{2}\log_2 x$

Vertical Dilation and Horizontal Translation

Equation (any positive base)	Transformation from $y = x$	Point Transformation
$y = k\log_2(x+b)$	Horizontal translation left b units then a Vertical dilation by a scale factor of k	$(x, y) \to (x-b, ky)$
$y = k\log_2(x-b)$	Horizontal translation right b units then a Vertical dilation by a scale factor of k	$(x, y) \to (x+b, ky)$

Example 63

Starting with $y = \log x$, apply these transformations. Translate horizontally 3 units right and apply a vertical dilation by a scale factor of $\frac{3}{4}$. What is the final equation?

Solutions:

Horizontal translation right 3 $\Rightarrow \log(x-3)$, Vertical dilation of scale factor $\frac{3}{4} \Rightarrow \frac{3}{4}\log(x-3)$

So, new equation $y = \dfrac{3}{4}\log(x-3)$

Exercise:

1. The point $(4, 2)$ lies on $y = \log_2(x)$. Find the corresponding point on $y = 3\log_2(x-1)$.
2. Write down the transformation required to change $y = \log_2 x$ to $y = 2\log_2(x+3)$.
3. The points $(1, 0)$, $(2, 1)$, and $(8, 3)$ lie on $y = \log_2 x$, find the corresponding points on $y = \frac{1}{2}\log_2(x+4)$.
4. Write the equation of the function that results when $y = \log(x)$ undergoes a horizontal translation 5 units left and a vertical dilation by a scale factor of 2.
5. The point $(29, 6)$ lies on the graph of $y = 2\log_3(x-2)$. What point lies on the parent function $y = \log_3 x$?
6. The function $y = \log x$ undergoes a vertical dilation and a horizontal translation to form a new function $y = 4\log(x+3)$.

   a State the horizontal transformation

   b State the vertical dilation factor

   c Find the new vertical asymptote

   d Find the new domain

   e Find the new x and y intercepts

7. The logarithmic function $y = \log_2 x$ has been transformed so that: Its vertical asymptote is at $x = -2$, It underwent a vertical dilation (stretched vertically) by a scale factor of 3. It passes through the point $(6, 9)$. Write the equation of this new function.

Vertical Dilation and Vertical translation

Order Matters!!!

Order matters when you have two vertical transformations

Equation (any positive base)	Transformation/Order from $y = \log_2 x$	Point Transformation
$y = k\log_2(x) + c$	Vertical dilation of scale factor k then Vertical translation up c units	$(x, y) \to (x, ky + c)$
$y = k(\log_2(x) - c)$	Vertical translation down c units then Vertical dilation of scale factor k	$(x, y) \to (x, k(y - c))$

Example 64

a Describe the transformations required to transform the graph of $y = \log(x)$ to $y = 2(\log(x)+3)$

b Describe 2 transformations required to transform the graph of $y = \log(x)$ to $y = 4\log(x) + 6$.

c The point $(4, 2)$ lies on the graph of $y = \log x$. What are the coordinates on the transformed graph $y = 4\log(x) + 6$?

Solutions:

a Vertical translation up 3 units then a vertical dilation of scale factor 2

b Vertical dilation of scale factor 4 then a vertical translation up 6 units
OR Vertical translation up $\frac{3}{2}$ units then vertical dilation of scale factor 4

c $(4, 2) \to (4, 2(4)) \to (4, 8 + 6) \to (4, 14)$

Exercise:

1. Given the graph of $y = \log_2 x$. Describe the transformation to obtain of the following
   a $y = 5\log(x) + 2$
   b $y = 5(\log(x) + 2)$
2. The point $(9, 2)$ lies on $y = \log_3 x$. Find the corresponding point on $y = 4\log_3(x) - 3$.
3. If $(8, 3)$ is on $y = \log_2 x$, find the corresponding point on $y = \frac{1}{3}(\log_2(x) + 3)$.
4. The graph of $y = \log_2 x$ is transformed to $y = 2\log_2(x) + 1$. Describe the transformations.
5. The function $y = \log x$ undergoes a vertical dilation of scale factor 2 followed by a vertical translation up 3 units. What is the equation of the new function?
6. The function $y = \log x$ undergoes a vertical translation up 3 units followed by a vertical dilation of scale factor 2. What is the equation of the new function?

Horizontal Dilation

Equation (any positive base)	Transformation from $y = \log_2 x$	Point Transformation
$y = \log_2(ax), a > 0$	Horizontal dilation of scale factor $\frac{1}{a}$	$(x, y) \rightarrow \left(\frac{1}{a}x, y\right)$

Note: We say there is a dilation in the x direction by a scale factor of $\frac{1}{a}$

or

We can also say there is a dilation from the y-axis by a scale factor of $\frac{1}{a}$

When $a > 1$, we say there is a horizontal compression by a scale factor of $\frac{1}{a}$

When $0 < a < 1$, we say there is a horizontal stretch by a scale factor of $\frac{1}{a}$

Example 65

If $y = \log_4(x)$ is transformed to the function $y = \log_4(2x)$

a Identify the horizontal dilation factor

b If $(4, 1)$ lies on $y = \log_4 x$, what is the corresponding point on $y = \log_4(2x)$

Solutions:

a the function has a horizontal dilation scale factor of $\frac{1}{2}$,

b $(4, 1) \rightarrow \left(\frac{4}{2}, 1\right) \rightarrow (2, 1)$

Exercise:

1. Given the graph of $y = \log_2 x$. Describe the transformation to obtain of the following

   a $y = \log_2 3x$

   b $y = \log_2 \frac{x}{3}$

2. The point $(9, 2)$ lies on $y = \log_3 x$. Find the corresponding point on $y = \log_3 3x$.

3. If $(16, 4)$ is on $y = \log_2 x$, find the corresponding point on $y = \log_2 \frac{x}{4}$.

4. The graph of $y = \log_3 x$ is transformed to $y = \log_3 5x$

   a Describe the horizontal transformation

   b State the vertical asymptote

   c State the domain

   d Find the new x intercept

Horizontal Dilation and Vertical Translation

Equation (any positive base)	Transformation from $y = \log_2 x$	Point Transformation
$y = \log_2(ax) + c, a > 0$	Horizontal dilation of scale factor $\frac{1}{a}$ and Vertical translation up c units	$(x, y) \to \left(\frac{1}{a}x, y + c\right)$
$y = \log_2(ax) - c, a > 0$	Horizontal dilation of scale factor $\frac{1}{a}$ and Vertical translation down c units	$(x, y) \to \left(\frac{1}{a}x, y - c\right)$

Example 66

If $(81, 4)$ lies on $y = \log_3 x$, find the corresponding point on $y = \log_3(3x) - 7$

Solutions:

$(81, 4) \to \left(\frac{81}{3}, 4 - 7\right) \to (27, -3)$

Exercise:

1. The function $y = \log_4 x$ is transformed by a horizontal dilation of scale factor 8 and then a vertical translation down 5 units. What is the resulting equation?

2. Write down the equation of the new function formed if the curve $y = \log_2 x$ is dilated horizontally by a scale factor of $\frac{1}{3}$ followed by a vertical translation up 2 units.

3. The point $(16, 4)$ lies on $y = \log_2 x$. Find the corresponding point on $y = \log_2\left(\frac{x}{4}\right) + 5$.

4. The function $y = \log_2 x$ is transformed to the function $y = \log_2\left(\frac{x}{3}\right) + 4$

   a Identify the horizontal dilation

   b Find the x intercept

5. Write the equation of a function that transforms $y = \log x$ such that the point $(100, 2)$ becomes $(20, 5)$ that involves only horizontal dilation and vertical shift

Vertical Translation, Vertical Dilation, and Horizontal Translation

Note: Remember for consistency, we will apply horizontal transformations before vertical transformations. The vertical dilation and vertical translation must be done in the correct order.

Equation (any positive base)	**Transformation/Order from $y = \log_2 x$**	**Point Transformation**
$y = k\log_2(x + b) + c$	Horizontal translation left b units then Vertical dilation of scale factor k then Vertical translation up c units	$(x, y) \to (x - b, ky + c)$

Example 67

The function $y = 3\log_2(x + 4) - 1$ is transformed to $y = \log_2(x - 2)$.

a What transformations occurred?

b If $(12, 11)$ lies on the original curve, what is the corresponding point on the new function?

c What transformations would be required to transform $y = \log_2(x - 2)$ to $y = 3\log_2(x + 4) - 1$

Solutions:

a The following transformations occurred: A horizontal translation 6 units to the right, then a vertical translation up 1 unit, then a vertical dilation by scale factor $\frac{1}{3}$

b $(12, 11) \to (12 + 6, (11 + 1) \div 3) \to (18, 4)$

c a horizontal translation left 6 units, a vertical dilation of factor 3 and a vertical translation down 1.

Example 68

The point $(9, 2)$ lies on $y = \log_3 x$. Find the corresponding point on $y = 3\log_3(x - 4) - 5$.

Solutions:

The transformation of $(9, 2) \to (9 + 4, 3(2) - 5) \to (13, 1)$

Exercise:

1. The point $(16, 4)$ lies on $y = \log_2 x$. Find the corresponding point on:

   a $y = -3\log_2(x + 2) + 1$

   b $y = 2\log_2(x - 3) + 5$

   c $y = \frac{1}{2}\log_2(x - 2) - 3$

   d $y = -2\log_2(x - 3) + 6$

2. Explain how you would transform the graph of $y = \log_2 x$ to obtain each of the functions in question 1.

3. Write the equation when $y = \log_3 x$ undergoes the following transformations:
   Horizontal translation 7 units right then a vertical dilation by scale factor 5 and finally a vertical translation 3 units down.

4. Write the equation when $y = \log_4(x - 1) + 2$ undergoes a horizontal translation 3 units left then a reflection about $x-$ axis, a vertical dilation by scale factor 2 followed by a vertical translation 5 units up.

5. The point $P(3, -4)$ lies on the graph of a function. It is transformed 2 units vertically up and then reflected in the $x-$ axis. What are the coordinates of P on the final function?

6. If $(64, 6)$ lies on $y = \log_2 x$, determine the corresponding point on $y = \frac{2}{3}\log_2(x - 20) + 3$.

7. A function of the form $y = 2\log_4(x + h) + k$ has its vertical asymptote at $x = -7$ and passes through $(9, 6)$. Find the value of h and k

Horizontal Dilation and Horizontal Translation

Order Matters!!!

Order matters when you have two horizontal transformations

Equation (any positive base)	Transformation/Order from $y = \log_2 x$	Point Transformation
$y = \log_2(ax + b), a > 0$	Horizontal translation left b units then Horizontal dilation of scale factor $\frac{1}{a}$	$(x, y) \to \left(\frac{1}{a}(x - b), y\right)$
$y = \log_2(a(x - b)), a > 0$	Horizontal dilation of scale factor $\frac{1}{a}$ then Horizontal translation right b units	$(x, y) \to \left(\frac{1}{a}x + b, y\right)$

Example 69

Find the new equation if $y = \log x$ is

a horizontally translated left 2 units and then horizontally dilated by a scale factor of $\frac{1}{3}$

Solutions:

$y = \log x \to y = \log(x + 2) \to y = \log(3x + 2)$

b horizontally dilated by a scale factor of $\frac{1}{3}$ then horizontally translated left 2 units

Solutions:

$y = \log x \to \log(3x) \to \log(3(x + 2))$

Example 70

Transform $y = \log x$ to create each function. State the order of operations clearly

a $f(x) = \log(2(x - 5))$

Solutions:

The transformations are
a horizontal dilation of scale factor $\frac{1}{2}$ then
a horizontal translation right 5

b $g(x) = \log(2x - 10)$

Solutions:

The transformations are
a horizontal translation right 10 then
a horizontal dilation of scale factor $\frac{1}{2}$

Exercise:

1. If $y = \log_2 x$ and $f(x) = \log_2(3(x-4))$.

   a Describe a sequence of transformations from y to $f(x)$

   b Find the $x-$ intercept of $f(x)$.

2. Consider the function $g(x) = \log_3(2x-8)$

   a Rewrite this function in factored form $g(x) = \log_3(a(x-h))$

   b Starting from $y = \log_3 x$, describe the order of transformations needed to obtain $g(x)$

3. Find the equation of the final graph when each of the following sequence of transformations have been applied to $y = \log x$

   a Translated right by two units then dilated horizontally by a scale factor of $\frac{1}{3}$

   b Dilated horizontally by a scale factor of $\frac{1}{3}$ then translated right by 2 units

   c If $\log x$ is dilated horizontally by a scale factor of $\frac{1}{3}$ first, what should be the translation scale factor to obtain the same answer as in part a?

4. If we start with $f(x) = \log_5 x$, write the equation after each step

   a Translated left by 2 units then dilated horizontally by a scale factor of 3.

   b Dilated horizontally by a scale factor of 3 then translated left by 2 units

   c If $f(x)$ is dilated horizontally by a scale factor of 3 first, what should be the translation scale factor to obtain the same answer as in part a?

5. Consider the functions $f(x) = \log(2(x-3))$ and $g(x) = \log(2x-3)$

   a Describe a sequence of transformation from $y = \log x$ to obtain $f(x)$

   b Describe a sequence of transformation from $y = \log x$ to obtain $g(x)$

   c Find the domain of each function.

   d Find the $x-$ intercept of each function

   e Find the vertical asymptote of each function.

6. The point $(9, 2)$ lies on $y = \log_3 x$. After a transformation what would be the corresponding point on the curve $y = \log_3(3x-6)$.

Horizontal Dilation, Horizontal Translation, Vertical Dilation, and Vertical Translation (Not Factorised)

Note: Order matters within horizontal transformations and within vertical transformations

Equation (any positive base)	Transformation/Order from $y = \log_2 x$	Point Transformation
$y = k\log_2(ax + b) + c$ $a > 0$	Horizontal translation left b units then Horizontal dilation of scale factor $\frac{1}{a}$ then Vertical dilation of scale factor k and Vertical translation up c units	$(x, y) \to \left(\frac{1}{a}(x - b), ky + c\right)$
$y = k\log_2(ax - b) - c$ $a > 0$	Horizontal translation right b units then Horizontal dilation of scale factor $\frac{1}{a}$ then Vertical dilation of scale factor k and Vertical translation down c units	$(x, y) \to \left(\frac{1}{a}(x + b), ky - c\right)$

Note: If $a < 0$, then there is a reflection about the y axis

If $k < 0$, then then there is a reflection about the x axis

Example 71

A transformation is applied to $y = \log(x)$ to create the function: $f(x) = -3\log(4x+20)+15$. Identify all the transformations from y to $f(x)$ in order.

Solutions:

A horizontal translation left 20.
a horizontal dilation scale factor of $\frac{1}{4}$
a vertical dilation scale factor of 3
reflection about the x-axis
a vertical translation up 15.

Exercise:

1. Start with $f(x) = -3\log(4x + 20) + 15$ and write a sequence of transformations to obtain $y = \log x$
2. The point $P(1, 0)$ lies on $y = \log_3(x)$. Find the coordinates of P after applying all transformations to obtain $h(x) = 2\log_3(6x + 18) - 5$
3. The graph of $y = \log_2 x$ is shifted 5 units right, reflected over the x axis and then shifted up 1 unit. What is equation of the new function?
4. The function $f(x) = \log_2(4x - 8) + 3$ is transformed from $g(x) = \log_2(x)$. Describe the required transformations in order.
5. The graph of $f(x) = \log_5(x)$ is transformed to create $g(x) = -2\log_5(3x + 12) + 7$. Describe, in order the transformations required.

Horizontal Dilation, Horizontal Translation, Vertical Dilation, and Vertical Translation (Factorised)

Note: Order matters within horizontal transformations and within vertical transformations. Remember for consistency, we will apply horizontal transformations before vertical transformations.

Equation (any positive base)	**Transformation/Order from $y = \log_2 x$**	**Point Transformation**
$y = k\log_2 a(x+b)+c$ $a > 0$	Horizontal dilation of scale factor $\frac{1}{a}$ then Horizontal translation left b units then Vertical dilation of scale factor k and Vertical translation up c units	$(x, y) \to \left(\frac{1}{a}x - b, ky + c\right)$
$y = k\log_2 a(x-b)-c$ $a > 0$	Horizontal dilation of scale factor $\frac{1}{a}$ then Horizontal translation right b units then Vertical dilation of scale factor k and Vertical translation down c units	$(x, y) \to \left(\frac{1}{a}x + b, ky - c\right)$

Note: If $a < 0$, there is a reflection about the y axis

If $k < 0$, there is a reflection about the x axis

Example 72

Apply the transformations to $y = \log x$ and write down the resultant function at each stage.
A horizontal dilation of scale factor $\frac{1}{2}$ then a vertical dilation by a scale factor of 4, a horizontal translation 3 units right and a vertical translation up 2 units.

Solutions:

$y = k\log(a(x-b)) + c$

A horizontal dilation of scale factor $\frac{1}{2} \to y = \log 2x$

then a vertical dilation by scale factor $4 \to y = 4\log 2x$

or a horizontal translation 3 units right $\to y = 4\log 2(x-3)$

followed by vertical translation 2 units up $\to y = 4\log 2(x-3) + 2$

Exercise:

1. Identify the transformations needed to transform $y = \log_2 x$ to $h(x) = \frac{1}{2}\log_2(4(x-1)) + 5$
2. Describe the transformations that map $y = \log_2 x$ to $y = 2\log_2(\frac{1}{4}(x-3)) + 1$
3. The point $P(8, 3)$ lies on $y = \log_2 x$. Find the coordinates of P after undergoing transformations that give $f(x) = \log_2(2(x-1)) + 4$
4. Given $f(x) = -3\log_2(2(x+1)) - 4$ and $y = \log_2 x$? State all the transformations needed to transform from $f(x)$ to y and from y to $f(x)$

RECAP

When does Order matter?

Order matters within horizontal transformations and within vertical transformations, but not between horizontal and vertical transformations

Order matters when you have

Horizontal Dilation (or reflection about the y-axis) AND a Horizontal Translation

Or

Vertical Dilation (or reflection about the x-axis) AND a Vertical Translation

Exercise:

State clearly if order matters for the following transformations	**Matter / Does not matter**
1. Translate horizontally and translate vertically	
2. Horizontally dilate and vertically dilate	
3. Horizontally dilate and translate horizontally	
4. Horizontally dilate and vertically translate	
5. Vertically dilate and vertically translate	
6. Vertically dilate and reflect about the x-axis	
7. Horizontally translate and reflect about the y-axis	
8. Horizontally translate and reflect about the x-axis	
9. Vertically dilate and reflect about the y-axis	
10. Vertically dilate and horizontally translate	
11. Reflect about the y-axis and reflect about the x-axis	
12. Horizontally dilate and reflect in the y-axis	

Graph Review

Describe Transformations given Graphs

1. List and explain, in order, the transformation needed to change the original function into the new function

a $\log x \to \log(x-1)$

b $\log x \to -1+\log x$

c $\log x \to \log 3x$

d $\log x \to 3\log 2x$

e $\log x \to \frac{1}{3}\log\frac{x}{2}$

f $\log x \to \log(4-x)$

g $\log x \to \log(x+1)-2$

h $\log x \to -\log(x-3)$

i $\log x \to 3\log(x+2)$

j $\log(x+1) \to \log(x-4)$

2. For each new function in question 1 above state the domain and asymptote.

3. Describe the transformations needed to change each given function back to $y = \log x$

a $y = \log 5x$

b $y = 3\log(x-2)+5$

c $y = -2\log(x+1)$

d $y = \log(3x-6)$

e $y = \log 3(x-6)$

Find Graphs Given the Transformations

4. Starting with the graph of $y = \log x$ write the equation of the function after applying the following transformations

a Dilate horizontally by a scale factor of 2 then translate right by 3 units

b Translate right by 3 units then dilate horizontally by a scale factor of 2

c Translate left by 1unit then dilate horizontally by a scale factor of 1/3

d Dilate horizontally by a scale factor of 1/3 third then translate left by 1 unit

e Translate left by 5 units then reflect in the y-axis

f Reflect in the y-axis then translate left by 5 units

5. Starting with the graph of $y = \log_2 x$ write the equation of the function after applying the following transformations

a Translate right by 3 units and then reflect across the y- axis

b Reflect about the y-axis and then translate right by 3 units

c Compare and comment on your results in a and b

6. Starting with the graph of $y = \log_2 x$ apply each transformation in the given order and write the equation after each stage

a $\log_2 x \xrightarrow[\text{right 2 units}]{\text{Horizontal translate}} \boxed{} \xrightarrow[\text{the } y\text{-axis}]{\text{Reflect across}} \boxed{} \xrightarrow[\text{by scale factor of 3}]{\text{Dilate Horizontally}} \boxed{}$

b $\log_2 x \xrightarrow[\text{right 2 units}]{\text{Horizontal translate}} \boxed{} \xrightarrow[\text{by scale factor of 3}]{\text{Dilate Horizontally}} \boxed{} \xrightarrow[\text{the } y\text{-axis}]{\text{Reflect across}} \boxed{}$

c $\log_2 x \xrightarrow[\text{the } y\text{-axis}]{\text{Reflect across}} \boxed{} \xrightarrow[\text{by scale factor of 3}]{\text{Dilate Horizontally}} \boxed{} \xrightarrow[\text{right 2 units}]{\text{Horizontal translate}} \boxed{}$

d $\log_2 x \xrightarrow[\text{the } y\text{-axis}]{\text{Reflect across}} \boxed{} \xrightarrow[\text{right 2 units}]{\text{Horizontal translate}} \boxed{} \xrightarrow[\text{by scale factor of 3}]{\text{Dilate Horizontally}} \boxed{}$

e $\log_2 x \xrightarrow[\text{by scale factor of 3}]{\text{Dilate Horizontally}} \boxed{} \xrightarrow[\text{right 2 units}]{\text{Horizontal translate}} \boxed{} \xrightarrow[\text{the } y\text{-axis}]{\text{Reflect across}} \boxed{}$

f $\log_2 x \xrightarrow[\text{by scale factor of 3}]{\text{Dilate Horizontally}} \boxed{} \xrightarrow[\text{the } y\text{-axis}]{\text{Reflect across}} \boxed{} \xrightarrow[\text{right 2 units}]{\text{Horizontal translate}} \boxed{}$

More Challenging Questions

7. The point $\left(15, 3\frac{1}{2}\right)$ lies on $y = \frac{1}{2}\log_5(x-10)+3$. What point lies on $y = \log_5(x+1)$?

8. List and explain, in order, the transformation needed to change the original function into the new function

a $\log(2x) \to \log(3-2x) - 1$

b $\log \frac{x}{3} \to 4\log(x) + 2$

c $\log(5x) - 3 \to \log(-2x) + 1$

9. Find two different sequences that transform $y = \log_2 x$ to $y = \log_2(2x-8)$

10. The graph $y = \log_2 x$ can be transformed to $y = \log_2(2-3x)$ using horizontal translation, reflection in the y-axis, and horizontal dilation. Describe three different orders in which these transformations can be applied.

11. Starting with function $y = \log_2(2-3x)$ and using the guide below, apply each transformation in sequence and write the resulting function at each stage until you reach $y = \log_2 x$

a $\log_2(2-3x) \xrightarrow[\text{the } y\text{-axis}]{\text{reflect across}} \boxed{} \xrightarrow[\text{a scale factor of 3}]{\text{Horizontally dilate by}} \boxed{} \xrightarrow[\text{right 2}]{\text{Horizontally Translate}} \boxed{}$

b $\log_2(2-3x) \xrightarrow[\text{left 2/3}]{\text{Horizontally Translate}} \boxed{} \xrightarrow[\text{the } y\text{-axis}]{\text{Reflect about}} \boxed{} \xrightarrow[\text{by scale factor of 3}]{\text{Horizontal Dilation}} \boxed{}$

Solutions

Page 1

1 $10^2 = 100$
2 $10^1 = 10$
3 $4^2 = 16$
4 $2^5 = 32$
5 $2^4 = 16$
6 $5^0 = 1$
7 $12^0 = 1$
8 $2^y = x$
9 $a^x = y$
10 $y^a = x$

Page 2

1 $\log_2 16 = 4$
2 $\log_3 9 = 2$
3 $\log_5 1 = 0$
4 $\log_4 64 = 3$
5 $\log_{10} 1000 = 3$
6 $\log_{10} 10 = 1$
7 $\log_7 7 = 1$
8 $\log_7 49 = 2$
9 $\log_7(\frac{1}{7}) = -1$
10 $\log_9 1 = 0$
11 $\log_5 \sqrt{5} = \frac{1}{2}$
12 $\log_3(\frac{1}{9}) = -2$
13 $\log_{10}(\frac{1}{10}) = -1$
14 $\log_5(\frac{1}{25}) = -2$

Page 3

1 2
2 3
3 5
4 3
5 4
6 3
7 3
8 7
9 4
10 4
11 $\frac{4}{3}$
12 $\frac{1}{2}$
13 2
14 $3\frac{1}{2}$

Page 4

1 -2
2 -2
3 -2
4 -1
5 -2
6 -5
7 -3
8 -2
9 $-\frac{3}{2}$
10 $-\frac{7}{4}$
11 $-\frac{3}{2}$
12 $-\frac{3}{2}$
13 $-\frac{5}{2}$
14 $-\frac{8}{3}$

Page 5

1 4
2 3
3 4
4 7
5 $\frac{1}{2}$
6 0
7 1
8 -1
9 $\frac{1}{2}$
10 $\frac{1}{25}$
11 5
12 81
13 1
14 12

Page 6

1 $2^{2x} = 2^{\frac{1}{2}}$, $x = \frac{1}{4}$
2 $3^{2x} = 3^{\frac{1}{2}}$, $x = \frac{1}{4}$
3 $5^x = 5^{-2}$, $x = -2$
4 $7^{2x} = 7^{\frac{1}{2}}$, $x = \frac{1}{4}$
5 $10^{2x} = 10^{\frac{1}{2}}$, $x = \frac{1}{4}$

Page 6 Continued

6 $2^{6x} = 2^{\frac{1}{2}}$, $x = \frac{1}{12}$
7 $3^{2x} = 3^{\frac{3}{2}}$, $x = \frac{3}{4}$
8 $2^{3x} = 2^{-2}$, $x = -\frac{2}{3}$
9 $2^{3x} = 2^{\frac{5}{2}}$, $x = \frac{5}{6}$
10 $5^{2x} = 5^{\frac{3}{2}}$, $x = \frac{3}{4}$
11 $x = \frac{1}{10}$
12 $x = 1000$
13 $x = \frac{1}{1000}$
14 $x = \frac{1}{2}$
15 $x = 4$
16 $x = \frac{1}{4}$
17 $x = \frac{1}{25}$
18 $x = \frac{1}{7}$
19 $x = \frac{1}{2}$
20 $x = \frac{1}{128}$

Page 7

1 $\frac{1}{2}$
2 4
3 0
4 $\frac{1}{3}$
5 $\frac{7}{2}$
6 -2

Page 8

1 1.362
2 1.973
3 0.845
4 -1.699
5 -0.319
6 2.966
7 $a = 1, b = 2$
8 $a = 0, b = 1$
9 $a = -1, b = 0$
10 $a = 2, b = 3$
11 $x = 3, y = 4$
12 $x = 2, y = 3$
13 $x = 5, y = 6$
14 $x = 3, y = 4$

Page 9

1 $x = 2$
2 $x = 2$
3 $x = 3$
4 $x = 10$
5 $x = \frac{1}{5}$
6 $x = 5$
7 $x = 2$
8 $x = 3$
9 $x = 8$
10 $x = 5$
11 $x = 10000$
12 $x = 16$
13 $x = 4$
14 $x = 3$

Page 10

1 1
2 1
3 1
4 1
5 1
6 1
7 1
8 1
9 0
10 0
11 0
12 0
13 0
14 0
15 0
16 0

Page 11

1 1
2 1
3 1
4 1
5 1
6 0
7 1
8 1
9 0
10 0
11 1

12 0
13 0
14 0

Page 12

1 1
2 1
3 1
4 1
5 1
6 1
7 1
8 1
9 1
10 1
11 1
12 1
13 0
14 0

Page 13

1 3
2 3
3 3
4 $\frac{1}{2}$
5 -1
6 $-\frac{1}{2}$
7 $\frac{5}{2}$
8 -2
9 $\frac{3}{2}$
10 -5

Page 14

1 2
2 3
3 2
4 4
5 1
6 $2 + \frac{1}{2}\log_2 5$
7 $4 + \frac{1}{2}\log_3 2$
8 $2 - \log_3 5$
9 $\log_2 7 - 3$
10 $2 - \log_5 9 = 2 - 2\log_5 3$

Page 15

1 $\log x + \log y$
2 $\frac{1}{2}(\log x + \log y)$
3 $\log x - \log y$
4 $\frac{1}{2}(\log x - \log y)$
5 $\log(x + y) - \log(x - y)$
6 $2\log(x + y)$
7 $-\frac{1}{2}\log(x + y)$
8 $\frac{1}{2}(\log(x + y) - \log(x - y))$
9 $\frac{1}{2}\log(x^2 + y^2)$
10 $2\log x + \frac{1}{2}\log y$
11 $2\log x + \log y + 3\log z$
12 $\log x - \log(y + z)$
13 $-(\log x + \log y + \log z)$
14 $\frac{1}{2}\log x + \log(y + z)$
15 $\log(x + y) - \frac{1}{2}\log(y + z)$
16 $\log(x^2 + y)$

Page 17

1 T
2 T
3 F
4 T
5 T
6 F
7 T
8 F
9 F
10 F
11 T
12 F
13 F
14 T
15 F
16 F
17 T
18 F
19 T
20 T
21 T
22 F

Page 18

1 $\log 2 + \log 3$
2 $1 + \log 2$
3 $1 + 2\log 3$
4 $3 + \log 2$
5 $\log 10 - \log 2 = 1 - \log 2$
6 $-\log 2$
7 $2(\log 2 + \log 3)$
8 $\frac{1}{2}(\log 3 + 2\log 2)$
9 $\log 2 - 1$
10 $\log 3 - \log 2$
11 $\log 64 - \log 100 = 6\log 2 - 2$
12 $2 - 2\log 2$
13 $\log 3 + 2 - 2\log 2$
14 $\log 3 - 2\log 2$
15 $3(\log 2 - \log 3)$
16 $\log\frac{18}{10} = \log 2 + 2\log 3 - 1$

Page 19

1 $\frac{1}{3}\log_7 8 = \frac{a}{3}$
2 $a - b$
3 $1 - a$
4 $-2a$
5 $a + b$
6 $1 + a$
7 $1 + b$
8 $-2b$
9 $\log_7 2 + \log_7 3 = \frac{1}{3}a + \frac{1}{2}b$
10 $\log_7 8 + \log_7 3 = a + \frac{1}{2}b$

Page 20

1 2.4
2 -1.2
3 0.6
4 2.2
5 3.2
6 $1 - 1.2 = -0.2$
7 $\frac{1}{2} - 1.2 = -0.7$
8 $1 - 1.2 = -0.2$
9 $\frac{1 + 1.2}{1.2 - 1} = \frac{2.2}{0.2} = 11$
10 $x = 1$
11 $x = 3$
12 $x = -1$
13 $x = -2$
14 $x = 0$
15 $x = \frac{5}{2}$

Page 21

1 $\log \frac{9 \times 6}{2} = \log 27$

2 $\log \frac{3}{6} = -\log 2$

3 $\log \frac{27}{9} = \log 3$

4 $\log \frac{20}{3}$

5 $\log \frac{7}{25}$

6 $\log \frac{16 \times 8}{2} = \log 64$

7 $\log \frac{9}{27} = -\log 3$

8 $\log xyz$

9 $\log \frac{x^3}{z^2}$

10 $\log \frac{xy \times yz}{xyz} = \log y$

11 $\log \frac{a^3}{a^2} = \log a$

12 $\log \frac{a(a+1)}{a^2} = \log \frac{a+1}{a}$

Page 22

1 3

2 $\frac{4}{3}$

3 $\frac{7}{3}$

4 3

5 3

6 3

7 4

8 2

9 $\frac{5}{7}$

10 $\frac{1}{4}$

11 $\frac{3}{2}$

12 $\frac{4}{3}$

13 2

14 2

Page 23

1 $x = 1$

2 $x = 2$

3 $x = 6$

4 $x = 8$

5 $x = 3$

6 $x = \frac{1}{2}$

7 $x = \frac{3}{2}$

8 $x = 16$

Page 24

1 $4x = 20$
$x = 5$

2 $\frac{x}{2} = 4$
$x = 8$

3 $x = 32$

4 $\frac{x}{x-2} = 2$
$x = 4$

5 $x + 1 = 2$
$x = 1$

6 $\frac{3x}{3-2x} = 6$
$x = \frac{6}{5}$

7 $3(x+1) = 9$
$x = 2$

8 $\log x = \log \frac{64}{4}$
$x = 16$

9 $\log x = \log(8 \times 10)$
$x = 80$

10 $\log_5 x = \log_5 \frac{9}{25}$
$x = \frac{9}{25}$

11 $x + 1 = 5 \times 100$
$x = 499$

12 $\frac{5+2x}{3} = x$
$x = 5$

Page 25

1 $(x-1)(x+3) = 5$
$x^2 + 2x - 8 = 0$
$x = 2, x = -4$ (not valid)

2 $(x-4)(x-3) = 6$
$(x-1)(x-6) = 0$
$x = 6, x = 1$ (not valid)

3 $(x+1)(x-2) = 10$
$x^2 - x - 12 = 0$
$x = 4, x = -3$ (not valid)

4 $x^2 - 1 = 8$
$x = 3, x = -3$ (not valid)

Page 25 continued

5 $(x-4)(x+2) = 7$
$(x-5)(x+3) = 0$
$x = 5, x = -3$ (not valid)

6 $x(x-1) = 12$
$(x-4)(x+3) = 0$
$x = 4, x = -3$ (not valid)

7 $2x(x+1) = 12$
$(x+3)(x-2) = 0$
$x = 2, x = -3$ (not valid)

8 $(\log_3 x)^2 - 4(\log_3 x) + 3 = 0$
$(\log_3 x - 3)(\log_3 x - 1) = 0$
$x = 27, x = 3$

9 $(\log x - 1)^2 = 0$
$\log x = 1$
$x = 10$

10 $\log_3 x = 2$ or $\log_3 x = -2$
$x = 9$ or $x = \frac{1}{9}$

Page 26

1 2.32

2 2.26

3 0.442

4 6.29

5 -1.43

6 $\frac{3}{2}$

7 $\frac{7}{2}$

8 -3

9 $\frac{5}{4}$

10 $\frac{3}{2}$

Page 27

1 $\frac{\log_2 x}{\log_2 \frac{1}{2}} = -b$

2 $\frac{\log_2 x}{\log_2 4} = \frac{b}{2}$

3 $\frac{\log_2 x}{\frac{1}{2}\log_2 2} = 2b$

4 $\frac{\log_2 x}{\log_2 2 + \log_2 x} = \frac{b}{1+b}$

5 $\frac{b}{1-b}$

6 $\frac{1}{1-b}$

7 $\log_x 4 - \log_x x = \dfrac{\log_2 4}{\log_2 x} - 1$

$= \dfrac{2}{b} - 1$

8 $\log_{\sqrt{x}} 2 + \log_{\sqrt{x}} x = \dfrac{\log_2 2}{\frac{1}{2}\log_2 x} + 2$

$= \dfrac{2}{b} + 2$

9 $\dfrac{\log_2 x}{\log_2 16} = \dfrac{b}{4}$

10 $\log_{16} 32 + \log_{16} x = \dfrac{5}{4} + \dfrac{b}{4}$

11 $\log_8 2 + \log_8 \sqrt{x} = \dfrac{1}{3} + \dfrac{\frac{1}{2}b}{3}$

$= \dfrac{1}{3} + \dfrac{b}{6}$

12 $\frac{1}{3} - \frac{b}{6}$

13 $\frac{2}{3}$

14 $\frac{7}{4}$

15 -3

16 $-\frac{4}{3}$

17 $-\frac{4}{3}$

18 3

19 1

20 3

21 $\log_x 3 = \dfrac{\log_x 9}{\log_x y}$

$\log_x 3 = \dfrac{2\log_x 3}{\log_x y}$

$\log_x y = 2$

$y = x^2$

22 $\log_x 5 = \dfrac{3\log_x 5}{\log_x y}$

$\log_x y = 3$

$y = x^3$

$\log_x y = \dfrac{\log_y y}{\log_y x}$

$= \dfrac{1}{\log_y x}$

23 $\log_{\sqrt{8}} 2 = \dfrac{1}{\log_2 \sqrt{8}}$

$= \dfrac{1}{\frac{3}{2}}$

$= \dfrac{2}{3}$

c $\dfrac{1}{\log_5 \sqrt{125}} = \frac{2}{3}$

Page 28

1 $x = \log_2 15$

$= 3.91$

2 $x = \log_5 7$

$= 1.21$

3 $x = \log_3 2$

$= 0.63$

4 $x = \log_9 8.9$

$= 0.99$

5 $x = \log_4 3.9$

$= 0.98$

6 $x = \log_5 10$

$= 1.43$

7 $x = \log_4 10$

$= 1.66$

8 $1 + 3x = \log_3 25$

$x = 0.64$

9 $2x = \log_3 10$

$x = 1.05$

10 $4x - 1 = \log_2 7$

$x = 0.95$

Page 29

1 $2^x \times 2^3 = 5^x$

$\left(\dfrac{5}{2}\right)^x = 8$

$x = 2.27$

or $(x+3)\log 2 = x \log 5$

$x\left(\log \dfrac{5}{2}\right) = 3\log 2$

$x = 2.27$

2 $\left(\dfrac{11}{5}\right)^x = 55$

$x = 5.08$

3 $\dfrac{4^{2x}}{4} = 5^x$

$\left(\dfrac{16}{5}\right)^x = 4$

$x = 1.19$

4 $\left(\dfrac{8}{5}\right)^x = 4 \times 125$

$x = 13.22$

5 $\left(\dfrac{144}{7}\right)^x = 7^5$

$x = 3.22$

6 $3^n = 37.5$

$n = 3.30$

7 $3^{2n} = 1.2$

$n = 0.08$

8 $4^x = 36$

$x = 2.58$

Page 30

1 $\log_5 x = 1$

$x = 5$

2 $\log_3 x = 5$

$x = 3^5 = 243$

3 $\log_2 x = 4$

$x = 16$

4 $\log_4 x = 16$

$x = 4^{16}$

5 $\log_3(\log_5 x) = 1$

$\log_5 x = 3$

$x = 5^3 = 125$

6 $\log_2(\log_4 x) = 1$

$\log_4 x = 2$

$x = 4^2 = 16$

7 $\log_2 8 = 3$

8 $\log_3 3 = 1$

9 $\log_5 5 = 1$

10 $\log_6(\log_3 3) = \log_6 1$

$= 0$

Page 31

1 4

2 1

3 $\frac{1}{5}$

4 $\sqrt{3}$

5 7

6 $\frac{1}{9}$

Page 32

1 T
2 T
3 F
4 F
5 F
6 F
7 T
8 T
9 F
10 F
11 F
12 T
13 T
14 F
15 F
16 F
17 T
18 F
19 F
20 T
21 T
22 T
23 T
24 F
25 F
26 F

Page 33

1 $10^{2.5} \approx 316$ times larger
2 1000 times more acidic
3a 1,000,000 times more than the library
b $1000 = 10^3 \rightarrow 3\text{x}10$
$= 30$ dB higher

Page 35

1a $2 = 2^{0.4t}$
$t = 2.5$ weeks
b $5 = 2^{0.4t}$
$t = 5.8$ weeks
2 $2500 = 1000 \times 3^{2k}$
$2.5 = 3^{2k}$
$k = \frac{1}{2}\log_3 2.5$

3 $2 = 1.08^t$
$t = \dfrac{\log 2}{\log 1.08}$
$= \log_{1.08} 2 \approx 9$ years

Page 36

4 $1.5 = 1.06^t$
$t = 6.96 \approx 7$ years
5 $\frac{50}{47} = 1.012^t$
$t = 5.19$ years
$= 5$yrs 2.5 months
So, 6 years
6a $t = 0, T = -15$
b $0 = -40(0.95)^t + 25$
$t \approx 9.2$min
$t \approx 10$minutes
c $20 = -40(0.95)^t + 25$
$0.95^t = 1/8$
$t \approx 41$ minutes
7a $2^t = \dfrac{P}{400}$
$t = \log_2(\dfrac{P}{400})$
b $t = \log_2 10 = 3.3$ weeks

Page 37

1 $x > \log_3 10$
$x > 2.10$
2 $x < \log_2 5$
$x < 2.32$
3 $x < \log_{0.4} 7$
$x < -2.12$
4 $x - 1 \leq \log_5 12$
$x \leq 2.54$
5 $x > \log_{0.8} 6$
$x > -8.03$
6 $x > 100$
7 $x \geq 9$
8 $0 < x < 8$
9 $0 < x \leq 25$
10 $0 < x - 1 \leq 8$
$1 < x \leq 9$

Page 38

1 3rd Line. $\log\frac{1}{7} < 0$
So the sign of the inequality should change
2 4th Line. $\log\frac{1}{2} < 0$
So the sign of the inequality should change
3 $\frac{1}{5} > \frac{1}{5}^3$
$\log\frac{1}{5} > 3\log\frac{1}{5}$
$1 > 3$

Page 39

1 $\begin{cases} \frac{x}{y} = 8 \\ x - 2y = 2 \end{cases}$
$y = \frac{1}{3}, x = \frac{8}{3}$
2 $\begin{cases} \frac{x}{y} = 4 \\ x + y = 8 \end{cases}$
$y = \frac{8}{5}, x = \frac{32}{5}$
3 $\begin{cases} y = 64 \\ y = 16x \end{cases}$
$y = 64, x = 4$
4 $\begin{cases} x = y^2 \\ x + 2y = 8 \end{cases}$
$y \neq -4, y = 2, x = 4$
5 $\begin{cases} xy^2 = 1 \\ x^2y = 27 \end{cases}$
$y = \frac{1}{3}, x = 9$
6 $20 = a\log_3 27b \quad (1)$
$-5 = a\log_3\frac{b}{9} \quad (2)$
$(1) - (2)$
$25 = 5a\log_3 3$
$b = 3, a = 5$
7 $\begin{cases} a = -\log_b 4 \\ 2 = a + \log_b 64 \end{cases}$
$b = 4, a = -1$

Page 40

1 $x > 3$ or $(3, \infty)$

2 $x > 0$ or $(0, \infty)$

3 $x > -3$ or $(-3, \infty)$

4 $x < 3$ or $(-\infty, 3)$

5 $x > 0$ or $(0, \infty)$

6 $x < 0$ or $(-\infty, 0)$

7 $x > 0$ or $(0, \infty)$

8 $x > -5$ or $(-5, \infty)$

9 All $x \in \mathbb{R}$ or $(-\infty, \infty)$

10 All $x \in \mathbb{R}, x \neq 0$ or $(-\infty, 0) \cup (0, \infty)$

Page 43
Review 1

1a 1
b 1
c 2
d 4
e 2
f -1
g $\frac{1}{2}$
h -2
i $-\frac{1}{2}$

2a 2
b 0
c 1
d 2
e 4
f 1
g 1
h 1
i 2
j 1
k 3
l 3

3a 1
b 0
c -1
d 3
e $\frac{1}{2}$
f $-\frac{1}{2}$
g $\frac{3}{2}$
h -2
i $\frac{5}{2}$
j $\frac{1}{3}$
k 2
l -2

4a $x = 9$
b $x = 8$

Page 44

c $x = 1$
d $x = 2^{-3} = \frac{1}{8}$
e $x = \frac{1}{5}$
f $x = 10$
g $x = \frac{1}{9}$

5a $x = 6$
b $x = -2$
c $x = 0$
d $x = -1$
e $x = 3$
f $x = \frac{5}{2}$
g $3x = \frac{1}{2}$
$x = \frac{1}{6}$
h $2x = \frac{3}{2}$
$x = \frac{3}{4}$
i $x = 5$
j $x = \frac{1}{4}$

6a $2 + x$
b $3x - 1$

7a 0.72
b 3.44

8a $\log_3 2$
b $\dfrac{4 \log 3}{3 \log 3} = \dfrac{4}{3}$
c $x^3 y^2$
d $\dfrac{5 \log_5 2}{2 \log_5 2} = \dfrac{5}{2}$

9a $3 \log x + \log y$
b $\dfrac{1}{2}(\log(x + y) - \log(x - y))$

10a $x > 1$ or $(1, \infty)$
b $x = 1$
c $0 < x < 1, (0, 1)$
d $x < 0$ or $(-\infty, 0)$

11a $\log_5 125 = 3$
b $\log_{64} 16 = \frac{2}{3}$
c $\log_a \sqrt{a} = \frac{1}{2}$
d $\log_a \frac{1}{a} = -1$

12a $2^5 = 32$
b $8^{\frac{2}{3}} = 4$
c $a^0 = 1$
d $a^1 = a$

13a $-\log_a m$
b $\frac{1}{2} \log_a m$
c x
d x

Page 45
Review 2

1a 3
b -3
c $-2\frac{1}{2}$
d $\frac{1}{3}$
e $\frac{1}{3}$
f -3

2a $x = \frac{5}{3}$
b $x = -2$
c $x = 4$
d $x = \frac{4}{5}$
e $x = -\frac{7}{8}$
f $x = 25$
g $x = 25$
h $x = \frac{1}{\sqrt{2}}$
i $x = 3$
j $x = 243$
k $x = \frac{1}{4}$
l $x = 6$

3a $x = 1$
b $x = 6$
c $x = 12$
d $x = 7$
e $x = \frac{1}{4}$
f $x = 1$
g $x = \frac{9}{8}$
h $x = 25$

4a $2 + 3 = 5$

b $\frac{1}{2} + \frac{1}{2} = 1$

c $\frac{3}{4}$

d $-1 - 2 = -3$

e $\frac{4}{3}$

f 3

g 6

5a $p + q$

b $r - p$

c $-p$

d $2r$

e $3q + 2p$

f $r - 2p$

g $r - 2p + 1$

h $2(r - p)$

i $\frac{3q}{2}$

Page 46

6a $2y$

b $-3y$

c $\frac{1}{2}y$

d $1 + \frac{1}{2}y$

e $2 + y$

7 $2^x = 5$
$8^x = 5^3 = 125$

8 $\log_{10} \frac{3}{100} = -2 + \log_{10} 3$

9a $\frac{7}{3}$

b $\frac{5}{2}$

10 $\log_5 15$

11a $1.6 - 1.2 = 0.4$

b $1.2 + 1.6 = 2.8$

c $\frac{1}{2}(1.2) = 0.6$

d $\log_x 3 + \log_x 2 = 0.6 + 0.4 = 1$

e $3 \log_x 3 = 1.8$

f $\log_x 9 - \log_x 2 = 1.2 - 0.4 = 0.8$

12 $\log_{10} 100 < \log_{10} 253 < \log_{10} 1000$

13 $\log_2 8 < \log_2 12 < \log_2 16$

14a $x = \log_7 11 = 1.232$

b $x + 1 = \log_4 49$
$x = 1.807$

Page 47
Review 3

1 $-2 - 1 + 0 + 1 + 2 = k$
$k = 0$ or
$\log_3 \frac{1}{9} \times \frac{1}{3} \times 1 \times 3 \times 9$
$= \log_3 1$
$= 0$

2a $\log x^3 y^7$

b $-6 \log x$

3a $\frac{3}{2}$

b 1

c 2

d 1

e $\frac{1}{2}$

f $\frac{1}{2}$

4a 3

b 3

c 1

5a $\dfrac{\log_x x}{\log_x y} = \dfrac{1}{\log_x y}$

b $\dfrac{\log_y x}{\log_y y} \times \dfrac{\log_y y}{\log_y x} = 1$

6a $\log_3 x = \log_3 \dfrac{3}{y}$
$x = \dfrac{3}{y}$

b $x = 8y$

c $x = \frac{81}{y^2}$

d $x = \frac{10}{y^2}$

7 Proof

8a $q - 1$

b $\log_{10} \frac{4}{100} = 2p - 2$

c $6p - 1$

9a $p = a^x$
$q = a^y$
$p^{\frac{1}{x}} = q^{\frac{1}{y}}$ or
$p^y = q^x$

b $x = \dfrac{\log p}{\log a}$
$y = \dfrac{\log q}{\log a}$
$\dfrac{\log p}{x} = \dfrac{\log q}{y}$
$y \log p = x \log q$
$p^y = q^x$

10 $\log 6 + \log 4 + \log 2 = \log 48$

11a $3^3 < 3^x < 3^4$
$27 < 30 < 81$

b $x = \log_3 30 = 3.0959 \approx 3.10$

Page 48

12a $5^x = 3$
$x = 0.68$

b $\left(\frac{81}{5}\right)^x = 125$
$x = 1.73$

c $\left(\frac{9}{5}\right)^x = 6$
$x = 3.05$

d $7^{2x} = 14$
$x = 0.68$

e $5^x = \frac{40}{3}$
$x = 1.61$

13a $x = \log_{10} 2$

b $x = \frac{1}{8}$

c $(\log_2 x - 4)(\log_2 x - 2) = 0$
$x = 16$ or $x = 4$

d $(3 \log_8 x - 4)(\log_8 x - 2) = 0$
$x = 16$ or $x = 64$

e $\log_3 x(\log_3 x - 5) = 0$
$x = 1$ or $x = 3^5 = 243$

14a $5000 = 3000(1.048)^t$
$t = 10.90$ (11 years)

b $5000 = 3000(1.012)^{4t}$
$t = 10.71$ years
= 10 years 9 months

15a $T = 6°C$

b $t = 2$ hours 47min

16a 1

b 6

c 9

17 $3^y = 7$
$9^y = 49$

18 $5x = \log_{10} \frac{y}{2}$
$x = \frac{1}{5} \log_{10} \frac{y}{2}$

19a $x < \log_5 12$
$x < 1.54$

b $x \log 0.5 > \log 3$
$x < \dfrac{\log 3}{\log 0.5} < -1.58$

c $0 < x < 32$

20 $0.5 = 0.9^t$
$t = 6.57$ (7th year)

Page 49
Extension

1 $\log_a 3 + \log_a 4 + \log_a 5 = \log_a 60$

2 $\log_{10} 5 = \dfrac{\log_3 5}{\log_3 10} = ab$

3a $\dfrac{2\log_3 13}{3\log_3 3} = \dfrac{2y}{3}$

b $\dfrac{4\log_5 3}{2\log_5 5} = 2y$

4 $\dfrac{2\log 3}{\log 2} \times \dfrac{3\log 2}{\log 5} \times \dfrac{-\log 5}{\log 3} = -6$

5a $(x+y)^2 = 25xy$
$\left(\dfrac{x+y}{5}\right)^2 = xy$

b $2\log\left(\dfrac{x+y}{5}\right) = \log xy$
$\log\left(\dfrac{x+y}{5}\right) = \dfrac{1}{2}(\log x + \log y)$

6 $\dfrac{\log_{10} 4 \times 3}{\log_{10} 5} = \dfrac{2p+q}{1-p}$

7 $\dfrac{2\log_2 15}{3\log_2 2} = p$
$p = \dfrac{2q}{3}$

8a $\log_3(\log_4 x) = 1$
$\log_4 x = 3$
$x = 64$

b $\log_2(\log_3 x) = 1$
$\log_3 x = 2$
$x = 9$

9 $2\log 2 - \log 3 = 2r - 2p - \dfrac{p}{2}$
$= 2r - \dfrac{5}{2}p$

10a $4^x = 36$
$x = 2.58$

b $\dfrac{\log 2}{\frac{1}{2}\log x} + \dfrac{\log x}{\log 4} = 2$
$(\log x - 2\log 2)^2 = 0$
$x = 4$

c $x = 9$

d $x = 2, x = 16$

11 $\dfrac{\log_a x}{\log_a ab} = \dfrac{\log_a x}{1 + \log_a b}$

12a $x \in \mathbb{R}, x \neq 0$

b $4 - x^2 > 0$
$-2 < x < 2$

c $x < -2$ or $x > 2$

d $x \in \mathbb{R}, x \neq 4$

e $-2 < x < 2$

f $x > 4$

13 $2\log x - \log y = 1$ and
$2\log x + 6\log y = 2$
subtract
$\log y = \dfrac{1}{7}, \log x = \dfrac{4}{7}$

a 4

b $\log x - \log y = \dfrac{3}{7}$

14 line 3, $\log \dfrac{1}{2} < 0$
so inequality sign should change

15a $\dfrac{\log_a x}{\log_a \frac{1}{a}} = \dfrac{\log_a x}{-1}$

b $\dfrac{\log_a a}{\log_a \frac{1}{a}} = \dfrac{1}{-1} = -1$

Page 51

1 $-3, -2, 1, 2, 3$

2 $-3, -2, 1, 2, 3$

3a Use graphics technology

b If $1 < a < 10$
Graph is shallower- not rising as rapidly but when $0 < a < 1$
Graph decreases less steeply

c Domain doesn't change Range doesn't change.

d No

e $\log 1 = 0$ so when $x = 1$, $y = 0$

Page 52

1 $y = 2 + \log x$

2 $y = -4 + \log x$

3 Translated down 3 units

4 $(10, 4)$

5 $\left(\dfrac{1}{10}, -3\right)$

6 $(10, 1)$

Page 53

1 $y = \log(x-2)$

2 $y = \log(x+4)$

3 Translated left 2 units

4 $(15, 1)$

5 $(9, 2)$

Page 54

1 $y = -1 + \log(x-2)$

2 $y = -3 + \log(x+4)$

3 Horizontally translated left 2 units and vertically down 3 units

4 $(96, 0)$

5 $(5, 1)$

Page 55

1a $(-9, 2)$

b $(9, -2)$

c $(-9, -2)$

2 Check on graphics software

3 $(64, -3)$

Page 56

1 $y = 2\log x$

2a Dilated vertically by a scale factor of 4

b $(1, 0)$

3 $(16, 20)$

4 $\left(\dfrac{1}{4}, -1\right)$

Page 57

1 $(5, 6)$

2 Translate horizontally left 3 units and dilated vertically by a scale factor of 2

3 $(-3,0)$ $(-2,\frac{1}{2})$
$(4,\frac{3}{2})$

4 $y = 2\log(x+5)$

5 $(27,3)$

6a 3 units left

b 4

c $x=-3$

d $x>-3$

e $(-2,0)$ $(0,4\log 3)$

7 $y = 3\log_2(x+2)$

Page 58

1a Vertical Dilation by a scale factor of 5 then vertically translated up 2

b Vertically translate up 2 then vertically dilate by a scale factor of 5

2 $(9,8)\to(9,5)$

3 $(8,6)\to(8,2)$

4 Vertically dilate by a scale factor of 2 then vertically up 1

5 $y = 2\log x + 3$

6 $y = 2(\log(x)+3)$

Page 59

1a Horizontally dilated by a scale factor of $\frac{1}{3}$

b Horizontally dilated by a scale factor of 3

2 $(3,2)$

3 $(64,4)$

4a Horizontally dilated by a scale factor of $\frac{1}{5}$

b $x=0$

c $x>0$

d $\left(\frac{1}{5},0\right)$

Page 60

1 $y = \log_4\left(\frac{x}{8}\right) - 5$

2 $y = 2 + \log_2 3x$

3 $(64,9)$

4a 3

b $\log_2\left(\frac{x}{3}\right) = -4$

$x = 2^{-4}\times 3 = \frac{3}{16}$

5 $y = \log(5x)+3$

Page 62

1a $(14,-11)$

b $(19,13)$

c $(18,-1)$

d $(19,-2)$

2a Translate left 2 units
Dilate Vertically 3 units
Reflexion in x-axis
Translate up 1 unit

b Translate right 3 units
Dilate vertically 2 units
Translate up 5 units

c Translate right 2 units
Dilate vertically $\frac{1}{2}$ units
Translate down 3 units

d Translate right 3 units
Reflexion in x-axis
Dilate Vertically 2 units
Translate up 6 units

3 $y = 5\log_3(x-7) - 3$

4 $\log_4(x+2)+2$
$\to -(\log_4(x+2)+2)$
$\to -2(\log_4(x+2)+2)$
$\to -2\log_4(x+2)+1$

5 $(3,-4)\to(3,-2)\to(3,2)$

6 $(84,7)$

7 $h=7, k=2$

Page 64

1a Dilate horizontally by a scale factor of $\frac{1}{3}$ then translate right 4 units

b $3(x-4)=1$

$x = \frac{13}{3}$

2a $y = \log_3 2(x-4)$

b Dilate horizontally by scale factor of $\frac{1}{2}$ then translate right 4 units

3a $y = \log(3x-2)$

b $y = \log 3(x-2)$

c $y = 3\left(x - \frac{2}{3}\right)$

4a $\log_5(x+2) \to \log_5\left(\frac{1}{3}x+2\right)$

b $\log_5\left(\frac{1}{3}x\right) \to \log_5\left(\frac{1}{3}(x+2)\right)$

c $\log_5\left(\frac{1}{3}x\right) \to \log_5\left(\frac{1}{3}(x+6)\right)$

5a Dilate horizontally by a scale factor of $\frac{1}{2}$ then translate horizontally right 3 units

b Translate horizontally right 3 units then dilate horizontally by a scale factor of $\frac{1}{2}$

c Domain $f(x)$: $x>3$
Domain $g(x)$: $x>\frac{3}{2}$

d $f(x)$: $2(x-3)=1$
$x = \frac{7}{2}$
$g(x)$: $2x-3=1$
$x=2$

e f(x): $x=3$
g(x): $x=\frac{3}{2}$

6 $(5,2)$

Page 65

1 Vertically translate down 15
Reflect in the x-axis
Vertically dilate by a scale factor of $\frac{1}{3}$
Dilate horizontally by a scale factor of $\frac{1}{4}$
Translate right 20 units

2 Translate horizontally left
$1-18\to-17$
Dilate horizontally by a scale factor of $\frac{1}{6}$ $\left(\frac{-17}{6}\right)$
Vertical dilation of scale factor 2. then a vertical translation down 5 units $0\to0\to-5$
New co-ordinates $(\frac{-17}{6},-5)$

3 $y = -\log_2(x-5)+1$

4 Horizontal translation right 8 units
Dilate horizontally of a scale factor of $\frac{1}{4}$.
Vertically translate up 3 units

5 Horizontal translation left 12 units
Dilate horizontally by a scale factor of $\dfrac{1}{3}$
Dilate vertically 2 units
Reflect about x-axis
Translate vertically up 7 units

Page 66

1 Dilate horizontally by a scale factor of $\frac{1}{4}$ then
Translate horizontally to the right 1 unit
Dilate vertically by a scale factor of $\frac{1}{2}$
Translate vertically up 5 units

2 Dilate horizontally by a scale factor of 4
Translation right 3 units
Dilate vertically by a scale factor of 2 units
Vertically translate up 1 unit

3 $(5, 7)$

4 $f(x) \to y$
Translate up 4 unit
Reflect about x-axis
Dilate vertically by a scale factor of $\dfrac{1}{3}$
Horizontally translate right 1 unit
Dilate horizontally by a scale factor of 2

$y \to f(x)$
Dilate horizontally by a scale factor of $\frac{1}{2}$
Translate horizontally left 1 unit
Dilate vertically by a scale factor of 3
Reflect about the x-axis
Translate down 4 units

Page 67

1 Doesn't Matter
2 Doesn't Matter
3 Matters
4 Doesn't Matter
5 Matter
6 Doesn't Matters
7 Matters
8 Doesn't Matter
9 Doesn't Matter
10 Doesn't Matter
11 Doesn't Matter
12 Doesn't Matter

Graph Review

1a Translate right 1 unit

b Translate down 1 unit

c Dilate horizontally by a scale factor of $\dfrac{1}{3}$

d Dilate horizontally by a scale factor of $\frac{1}{2}$
Dilate vertically by a scale factor of 3

e Dilate horizontally by a scale factor of 2
Dilate vertically by a scale factor of $\dfrac{1}{3}$

f Reflect about y-axis
Translate right 4 units
OR
Translate left 4 units
Reflect about the y-axis

g Translate left 1 unit
Translate down 2 units

h Translate right 3 units
Reflect about x-axis

i Translate left 2
Vertically dilate by scale factor of 3

j Translate right 5 units

2a $x > 1$ or $(1, \infty)$ $x = 1$
b $x > 0$ or $(0, \infty)$ $x = 0$
c $x > 0$ or $(0, \infty)$ $x = 0$
d $x > 0$ or $(0, \infty)$ $x = 0$
e $x > 0$ or $(0, \infty)$ $x = 0$
f $x < 4$ or $(-\infty, 4)$ $x = 4$
g $x > -1$ or $(-1, \infty)$ $x = -1$
h $x > 3$ or $(3, \infty)$ $x = 3$
i $x > -2$ or $(-2, \infty)$ $x = -2$
j $x > 4$ or $(4, \infty)$ $x = 4$

3a Dilate horizontally by a scale factor of 5

b Translate down 5 units
Dilate vertically by a scale factor of $\dfrac{1}{3}$
Translate horizontally left 2 units

c Reflect about the x-axis
Dilate vertically by a scale factor of $\frac{1}{2}$
Translate horizontally right 1 unit

d Dilate horizontally by a scale factor of 3
Translate horizontally left 6 units

e Translate horizontally left 6 units
Dilate horizontally by a scale factor of 3

4a $y = \log \frac{1}{2}(x-3)$

b $y = \log\left(\frac{1}{2}x - 3\right)$

c $y = \log(3x+1)$

d $y = \log 3(x+1)$

e $y = \log(-x+5)$

f $y = \log(-(x+5))$

5a $y = \log_2(-x-3)$

b $y = \log_2(-(x-3))$

c Orders matters

6a $\log_2(x-2) \to \log_2(-x-2) \to \log_2\left(-\frac{1}{3}x - 2\right)$

b $\log_2(x-2) \to \log_2\left(\frac{1}{3}x - 2\right) \to \log_2\left(-\frac{1}{3}x - 2\right)$

c $\log_2(-x) \to \log_2\left(-\frac{1}{3}x\right) \to \log_2\left(-\frac{1}{3}(x-2)\right)$

d $\log_2(-x) \to \log_2(-(x-2)) \to \log_2\left(-\left(\frac{1}{3}x - 2\right)\right)$

e $\log_2\left(\frac{1}{3}x\right) \to \log_2\frac{1}{3}(x-2) \to \log_2\left(-\frac{1}{3}(x+2)\right)$

f $\log_2\frac{1}{3}x \to \log_2\left(-\frac{1}{3}x\right) \to \log_2\left(-\frac{1}{3}(x-2)\right)$

7 $(4, 1)$

8a Reflect about the y-axis
Translate horizontally right $\frac{3}{2}$
Translate vertically down 1 unit

b Dilate horizontally by a scale factor of $\frac{1}{3}$
Dilate vertically by a scale factor of 4
Translate vertically up 2 unit

c Translate vertically up 4 units
Dilate horizontally by a scale factor of $\frac{5}{2}$
Reflect about the y-axis

9 Translate horizontally right 8 units
Dilate horizontally by a scale factor of $\frac{1}{2}$

OR

Dilate horizontally by a scale factor of $\frac{1}{2}$
Translate horizontally right 4 units

10

Reflect about the y-axis:

$y = \log_2(-x)$

Translate right 2 units:

$y = \log_2(-(x-2))$

Dilate horizontally by $\frac{1}{3}$:

$y = \log_2(-3x+2)$

OR

Translate left 2 units:

$y = \log_2(x+2)$

Dilate horizontally by $\frac{1}{3}$:

$y = \log_2(3x+2)$

Reflect about the y-axis:

$y = \log_2(-3x+2)$

OR

Dilate horizontally by $\frac{1}{3}$:

$y = \log_2(3x)$

Reflect about the y-axis:

$y = \log_2(-3x)$

Translate right $\frac{2}{3}$ units:

$y = \log_2\left(-3\left(x - \frac{2}{3}\right)\right)$

11a $\log_2(2+3x) \to \log_2(2+x) \to \log_2 x$

b $\log_2\left(2 - 3\left(x + \frac{2}{3}\right)\right) \to \log_2(-3(-x)) \to \log_2 3\left(\frac{x}{3}\right) = \log_2 x$

Made in the USA
Coppell, TX
06 November 2026

11224703R00052